T0262964

# LIPASE

An Industrial Enzyme Through Metagenomics

# LIPASE

An Industrial Enzyme Through Metagenomics

**B. K. Konwar, PhD**
**Kalpana Sagar, PhD**

Apple Academic Press Inc.              Apple Academic Press Inc.
3333 Mistwell Crescent                 9 Spinnaker Way
Oakville, ON L6L 0A2 Canada            Waretown, NJ 08758 USA

© 2018 by Apple Academic Press, Inc.

First issued in paperback 2021

*Exclusive worldwide distribution by CRC Press, a member of Taylor & Francis Group*
No claim to original U.S. Government works
ISBN 13: 978-1-77-463060-0 (pbk)
ISBN 13: 978-1-77-188618-5 (hbk)

### Library and Archives Canada Cataloguing in Publication

Konwar, B. K., author
Lipase : an industrial enzyme through metagenomics / Prof. B.K. Konwar, Dr. Kalpana Sagar.
Includes bibliographical references and index.
Issued in print and electronic formats.
ISBN 978-1-77188-618-5 (hardcover).--ISBN 978-1-315-15923-2 (PDF)

1. Lipase.  2. Metagenomics.  I. Sagar, Kalpana, author  II. Title.

| QP609.L5K66 2018 | 572'.757 | C2018-900086-4 | C2018-900087-2 |

### Library of Congress Cataloging-in-Publication Data

Names: Konwar, B. K., author. | Sagar, Kalpana, author.
Title: Lipase : an industrial enzyme through metagenomics / Prof. B. K. Konwar, Dr. Kalpana Sagar.
Description: Toronto ; New Jersey : Apple Academic Press, 2018. | Includes bibliographical references and index.
Identifiers: LCCN 2017061343 (print) | LCCN 2017061844 (ebook) | ISBN 9781315159232 (ebook) | ISBN 9781771886185 (hardcover : alk. paper)
Subjects: | MESH: Lipase--genetics | Metagenomics--methods
Classification: LCC QP609.L5 (ebook) | LCC QP609.L5 (print) | NLM QU 136 | DDC 572/.757--dc23
LC record available at https://lccn.loc.gov/2017061343

Apple Academic Press also publishes its books in a variety of electronic formats. Some content that appears in print may not be available in electronic format. For information about Apple Academic Press products, visit our website at **www.appleacademicpress.com** and the CRC Press website at **www.crcpress.com**

# CONTENTS

# PREFACE

Metagenomics, a molecular method based on direct isolation and analysis of nucleic acids, proteins and lipids from environmental samples, reveals structural and functional information about microbial communities. Microbial lipases are industrially important and have gained attention due to their stability, selectivity and broad substrate specificity. Bacteria, yeast and fungi are potential sources of lipases. The industrial demand for new sources of lipases, with different catalytic characteristics, stimulates the isolation and selection of new strains. Lipase-producing bacteria have been found in different habitats such as industrial wastes, vegetable oil processing factories, dairy plants and soil contaminated with oil and oil seeds among others. Among the various kinds of bacteria, *Bacillus* group exhibited interesting properties that make them potential candidates for biotechnological applications. Lipases are classified into eight families, I to VIII, based on their amino acid sequences. It is a widely accepted fact that more than 99% of microorganisms present in soil samples are not readily culturable and therefore not accessible for biotechnology or basic research. The genomes of these uncultured species encode a largely untapped reservoir of novel enzymes and metabolic capabilities. As such more than 99% of microorganisms have unique and potentially very useful abilities such as waste degradation, enzyme production and synthesis of compounds that could find use as drugs or antibiotics.

The present investigation was undertaken to explore the lipase-producing bacterial gene(s) in bakery soil. Metagenomic DNA (mgDNA) was extracted using Porteous et al. method with some modifications. The purity and cloneability of extracted mgDNA was confirmed by polymerase chain reaction (PCR) and restriction digestion. The isolated and purified soil mgDNA was partially digested with Sau3AI, and the DNA fragments 0.5–3.0 kb in size were ligated into *Bam*HI digested pET-32a vector. The vector was transferred to *Escherichia coli* BL21(DE3) pLysS competent cells. The recombinant clones were screened, and a positive clone carrying the recombinant plasmid KBS-plip1 exhibited lipolytic activity. The positive clone was confirmed with the band shift of the recombinant plasmid and DNA sequencing. The sequencing of the 996-bp insert DNA in the clone KBS-plip1 revealed an open reading frame (ORF) of 891 bp encoding the protein with 296 amino acids which showed 99% sequence homology with the DNA of the lipase/esterase-producing unculturable bacteria, KC182797. The phylogenetic tree was constructed using neighbour joining method. Subsequently, the nucleotide sequence was deposited in the BanKit of NCBI under the accession number KF743145.

Multiple sequence alignment of amino acids revealed 23–55% homology with the unculturable bacterial lipases. KBS-plip1 was classified as family IV or hormone-sensitive lipase (HSL) family. The HSL family conserved HGGG motif (amino acid 23–26) was found upstream of the active-site motif in KBS-plip1. Hormone-sensitive lipase contains the lipase conserved catalytic consensus pentapeptide GDSAG. Catalytic triad DPM (amino acids 145–147) and HVF (amino acids 228–230) are also the conserved sequences of HSL family and present in the KBS-plip1 amino acid sequence. The cloned lipase gene overexpressed in the presence of 0.5 mM Isopropyl β-D-1-thiogalactopyranoside (IPTG). The size of recombinant protein, fused with N- terminal His- tag, was found to be ~40.86 kDa including the insert sequence encoding 296 amino acids (33867.4 Da) and pET-32a vector backbone sequence contributing 55 amino acids (6988.52 Da). The modelled enzyme is a folded dimmer and consists of nine α-strand attachment with seven β-helices. The Ramachandran plot validated the modelled structure of KBS-plip1. This is significant as it revealed a good quality protein as high percentage ($>95\%$) of amino acid residues are in the favoured region.

Biochemical characterization of the KBS-plip1 lipase revealed maximum activity at a concentration of 1.0 $\mu g \cdot ml^{-1}$ and in the presence of 1.0% tributyrin, beyondthis the enzyme activity became saturated. The kinetic study of the KBS-plip1 lipase revealed $V_{max}$ and $K_m$ values to be 227 U/ml and 0.806 mg/ml, respectively. The purified protein exhibited a specific activity 96.32 U/mg with 2.23 purification fold. For maximum activity of the enzyme, a temperature of 37°C and pH 7.5, presence of the divalent cations $Ca^{2+}$, $Mn^{2+}$, $Zn^{2+}$ and $Fe^{2+}$ were essential whereas EDTA inhibited the catalytic activity; but CTAB and gum arabic enhanced the activity. NaCl at a concentration of 1.5 M increased the enzyme activity by 2.5 fold. The organic solvents such as ethanol, 1-propanol, acetone, acetonitrile, glycerol and DMSO enhanced the activity by 3.5 fold. The enzyme was stable at 37°C up to 100 min, but at 45°C and beyond (55°C) the enzyme activity decreased drastically with half-life of 60 and 20 min, respectively.

The thermal stability of the crude lipases isolated from the culturable soil bacterial strains KBS-101, KB2F, KBS-103, KBS-105 and KBS-107 retained 50% catalytic activity at 60°C up to 80 min. The enzyme demonstrated industrial application in laundry detergent formulation. On the basis of bacterial growth kinetics, lipase production yield, detergent stability and thermal stability of the lipases isolated from the bacterial strains KBS-101, KB2F, KBS-103, KBS-105 and KBS-107 were characterized. The 16S rDNA isolated from these bacterial isolates were amplified and sequenced. Phylogenetic and sequence homology study for the sequences of the bacterial strain KBS-101 showed 97% 16S rRNA sequence homology with *Bacillus* methylotrophicus strain (HQ844459) and represented to be its closest phylogenetic neighbour. *Bacillus tequilensis* strain (HQ234273) showed 99% sequence homology with KB2F, strain KBS-103 showed 99% sequence homology with the unculturable strain HQ764973. The strain KBS-105 showed 100% sequence homology with *Bacillus badius* strain GQ497939, KBS-10797% with *Enterobacter* sp. FJ440555.

An inoculum size of 2.0% was effective in producing maximum lipase by all the bacterial strains. The strains KBS-101 and KBS-105 possessed higher accumulation of lipase on supplementing the media with glucose as the carbon source, KB2F revealed the maximum accumulation in the presence of lactose, KBS-103 and KBS-107 in the presence of galactose. The maximum enzyme activity was shown by the strains KBS-101 and KB2F in the presence of beef extract, whereas KBS-103 and KBS-107 in the presence of yeast extract. The strain KBS-105 possessed higher accumulation of lipase when the medium was supplemented with peptone. The maximum substrate specificity for olive oil was revealed by the lipases isolated from the strains KBS-101, KBS-103 and KBS-105, whereas KB2F and KBS-107 showed for tributyrin. Maximum lipase accumulation and cell biomass was obtained in the case of KBS-101 and KBS-103 after 24 h, strains KB2F, KBS-105 and KBS-107 showed the maximum accumulation after 48 h. The optimum pH for maximum lipase accumulation by the bacterial strains KBS-101 and KBS-105 was 8.5, and for KB2F and KBS-103 it was 7.5 and KBS-107 at 9.0. The optimum temperature for maximum lipase accumulation from the bacterial strains KBS-101, KBS-103, KBS-105 and KBS-107 was 37°C and for KB2F it was 45°C. The agitation rate of 200 rpm revealed maximum lipase accumulation by all five bacterial strains. Out of divalent cations used, $Ca^{2+}$ enhanced the enzyme activity in the case of strains KBS-10 and KBS-103; $Fe^{2+}$ in the case of KB2F and KBS-107; $Co^{2+}$ in the case of KBS-105 and $Mg^{2+}$ in the case of KBS-107.

The culture supernatant containing crude lipases extracted from the bacterial strains exhibited antimicrobial activity against the bacterial species *E. coli* (MTCC 40), *Bacillus subtilis* (MTCC 619), *Staphylococcus aureus* (MTCC 737), *Klebsiella pneumoniae* (MTCC 109) and *Pseudomonas aeruginosa*; and two fungal species Candida *albicans* (3017) and *Fusarium oxysporum* (MTCC 284). The culture supernatant containing crude lipases from the bacterial strain KBS-101 showed antibacterial activity against all the bacterial strains; KBS-103 showed activity against *E. coli* (MTCC 40), *B. subtilis* (MTCC 619) and *S. aureus* (MTCC 737); whereas KBS-105 showed activity against *E. coli* (MTCC 40). The culture supernatant containing crude lipases of KBS-101 and KBS-105 exhibited the antifungal activity against *F. oxysporum* (MTCC 284).

This book is a humble effort of the authors to document the scientific findings carried out over the last 8 years in the area of unculturable soil microorganisms. The authors deeply acknowledge help and support rendered by Mr. Salam Pradeep Singh, Mr. Yasir Bashir and Dr. Mayur Mausoom Phukan. Authors would also like to sincerely thank Dr. (Mrs.) Juri G. Konwar, faculty member of Department of Cultural Studies, Tezpur University, Napaam, Assam, India; Ms. Chandrika Konwar, MSc student at Dr. B. R. Ambedkar Center for Biomedical Research, University of Delhi, North campus, India; and Mr. Chandrim Konwar, B. Tech student (computer science) at Amity University, Noida, India.

The book would not have seen the light without the willful approach of bright and dynamic Dr. Sarika Garg, CEO and Founder, HS Counseling, Canada representing Apple Academic Press (AAP). The authors owe a special thanks to her and the Apple Academic Press.

Napaam, Tezpur

**Dr. B. K. Konwar**
bkkon@tezu.ernet.in
Former Vice-Chancellor
Nagaland University (Central)

**Dr. Kalpana Sagar**
kalpsaga@gmail.com

# ABOUT THE AUTHORS

**B. K. Konwar, PhD**, is currently Professor in the Department of Molecular Biology and Biotechechnology at Tezpur University (Central), Assam, India. He obtained his MSc agricultural degree in plant breeding and genetics from Assam Agricultural University, Jorhat, Assam, India. He has worked as a lecturer, assistant professor, and associate professor at the university. His PhD in plant biotechnology is from the Imperial College of Science, Technology and Medicine, University of London, United Kingdom. He was formerly affiliated with the Tocklai Experimental Station, Tea Research Association, Jorhat, Assam, India, as a Senior Scientist (Botany and Biotechnology). Other appointments include Professor and Department Head, Molecular Biology and Biotechnology at Tezpur University (Central), Napaam, Tezpur, Assam, India; Head of the Centre for Petroleum Biotechnology, DBT (DST, Govt of India; Dean, School of Science and Technology, Tezpur University; and Vice-Chancellor, Nagaland University (Central), HQ: Lumami (campuses: Meriema, Medziphema, Dimapur), Nagaland, India.

Dr. Konwar has carried out 12 research projects of national importance as the principal investigator and has supervised over 40 MSc students and 14 PhD scholars who carried out research projects at Assam Agricultural University and Tezpur University. He and his research group so far have deposited 11 gene (DNA) sequences in gene banks and have published 142 research papers in reputed (IF: 1-6) international/national journals, along with 61 papers in seminar and conference proceedings and 74 research paper abstracts in India and abroad. He published more than 130 popular science, environment, biotechnology, history, national integration, higher education, research needs, and other articles in various Assamese magazines and newspapers, as well as more than 40 scientific articles in English magazines in addition to several books, booklets, and book chapters.

**Kalpana Sagar, PhD,** is currently a Research Associate at Delhi University. She has earned her PhD in molecular biology and biotechnology from Tezpur University under the supervision of Prof. B. K. Konwar.

# CHAPTER 1

# INTRODUCTION

## CONTENTS

Microorganisms inhabited this Earth before any other higher organism. During the course of evolution, they have occupied many diverse environments including soil, fresh and marine water and so forth. They also form symbiotic relationship with plants and animals including human beings. Microbes are the first living organisms and play a key role in global nutrient cycles such as the carbon, nitrogen and sulphur cycles and are valuable source of biodiversity. About 3.5 billion years ago, when life originated on the earth, it was solely represented by microbes. Microbes of untold diversity can be found in almost every habitat present in the nature including rather hostile environments, living typically in complex communities with different kinds of associations. Indeed, they can be found almost ubiquitously on earth be it commonly seen environments such as soil, air and water or the most extreme conditions such as volcano vents, Antarctic environments and deep sea hydrothermal vents or rocks in deep bore holes beneath the earth's surface. The relationship between environmental factors and ecosystem function can be understood by studying the bacterial community structure and diversity. The major goal of microbial ecology is to understand microbial community structure and diversity in natural habitats along with their interaction with one another and with their habitat. In addition to their essential activities throughout the biosphere, microbes have been the source of numerous technologies that have improved human conditions. They are also used commercially for the following:

1.  Production of most of the antibiotics and many other drugs in clinical use
2.  Remediation of pollutants in soil and water
3.  Enhancement of crop productivity
4.  Production of biofuels that are required today
5.  Fermentation of various edible items, and to provide unique signatures that form the basis of microbial detection in disease diagnosis and forensic analysis
6.  Mining waste waters, no matter how toxic the contaminants are for humans and animals
7.  Degradation of herbicides and pesticides, thus, cleaning up the groundwater
8.  Making ammonia, a component of dung and fertilizer, available to plants
9.  Specific bacterial species are also available for cleaning up industrial wastes, for example *Deinococcus radiodurans,* is the only bacteria so far known to survive high doses of radioactivity, hence, can be used to clean up radioactive wastes

There are bacteria which can degrade plastics as well as others which can even synthesize plastics. However, the major attributes associated with the microbial world are their potential usage in industry to obtain numerous products such as antibiotics, proteins, enzymes; their applications in agriculture, alcohol industry, pharmaceuticals, steroids, vaccines, vitamins, organic acids, amino acids, organic solvents, synthetic fuels, bioleaching, recovery of petrol, single-cell protein production and many more, the list being endless.

Besides antibiotics, enzymes are bioactive compounds that have focused attention. The demand of microbial enzymes is increasing rapidly due to their application in various industries such as food, pharmaceuticals, detergents, cosmetics, textiles, paper, leather and environment-friendly cost-effective biotechnological processes. The microbial enzymes act as biocatalysts to perform reactions in bioprocesses in an economical and environment friendly way as opposed to the use of chemical catalysts. The special characteristics of microbial enzymes are exploited for commercial interests and industrial applications. The special characteristics of microbial enzymes are exploited for commercial interests and industrial applications. These characteristics include thermophilic nature, tolerance to a varied range of pH, stability of enzyme activity over a range of temperature and pH, and other harsh reaction conditions. Such enzymes have proven their utility in bioconversions and bioremediation bioindustries. As per the above-mentioned exploitations, exploration of microbial enzymes and their biocatalytic machinery has been attracting a greater attention.

Current estimates indicate that majority of microorganisms present in many natural environments remain largely untapped, unknown and uncharacterized due to lack of proper culture conditions, therefore not accessible for biotechnology or basic research (Schloss and Handelsman, 2003; Riesenfeld et al., 2004). On the other hand, metagenomics represents a powerful tool to characterize microbial diversity present in native environmental samples regardless of the availability of laboratory culturing techniques. During the last two decades, development of methods to isolate nucleic acids from environmental sources has opened a window to previously unknown diversity of microorganisms. Analysis of nucleic acids directly extracted from environmental samples allows the study of natural microbial communities without the need for cultivation. Therefore, the taxonomy based on phylogeny, in which data from uncultured bacteria are included, is rapidly changing and replacing the former taxonomy based exclusively on morphological, physiological and biochemical parameters of cultured bacteria (Garrity et al., 2005; Brady et al., 2002; Frost and Sullivan, 2010). Each organism in an environment has a unique set of genes in its genome; the combined genomes of all the community members make up the "metagenome". Metagenomics has led to the accumulation of DNA sequences, and these sequences are exploited for novel biotechnological applications. Due to the overwhelming majority of unculturable microbes in soil, metagenome searches will always result in identification of hitherto unknown genes and proteins. Thus, the probability of uncovering hitherto unknown sequence makes this approach more favourable than searches in already cultured microbes.

## 1.1   ENZYMES

Enzymes are large biological molecules responsible for thousands of the chemical interconversions that sustain life. Enzymes are highly selective catalysts; they accelerate both the rate and specificity of metabolic reactions greatly from the digestion of food to the synthesis of DNA. Most enzymes are proteins and adopt a specific three-dimensional structure and may employ organic and inorganic cofactors to assist in catalysis. Almost all chemical reactions in a biological cell require enzymes in order to occur at rates sufficient for life. Like all catalysts, enzymes work by lowering the activation energy $(E_a^{\ddagger})$ for a reaction, thus dramatically increasing the rate of the reaction. As a result, products are formed faster and reactions reach their equilibrium state more rapidly. All catalysts are neither consumed by the reactions they catalyse nor do they alter the equilibrium of these reactions. Enzymes are known to catalyse about 4000 biochemical reactions. Enzyme activity can be affected by other molecules called inhibitors and activators. Inhibitors are molecules that decrease enzyme activity, and activators are molecules that increase enzyme activity. The activity of enzyme is also affected by temperature, pressure, chemical environment (e.g. pH) and the concentration of the substrate.

## 1.2   ETYMOLOGY AND HISTORICAL ASPECTS OF INDUSTRIAL ENZYMES

The history of enzyme technology began in 1874, when Christian Hansen, a Danish chemist, produced the first specimen of rennet by extracting dried calves' stomachs with saline solution, which was the first enzyme used for industrial purposes. The fermentative activity of microorganisms was discovered in 18th century by the French scientist Louis Pasteur. The term enzyme comes from *zymosis,* the Greek word for fermentation, a process accomplished by yeast cells and long known to the brewing industry, which occupied the attention of many 19th-century chemists.

In 1833, French chemist Anselme Payen discovered the first enzyme, diastase. A few decades later, while studying the fermentation of sugar to alcohol by yeast, Louis Pasteur came to the conclusion that the fermentation was catalysed by vital force contained within the yeast cells called ferments functions only within living organisms. In 1860, he recognized that enzymes were essential for fermentation but assumed that their catalytic action was inextricably linked to the structure and life of the yeast cell. In 1878, German physiologist Wilhelm Kuhne coined the term *enzyme* from Latin words, which literally means "in yeast". In 1897, German chemist Edward Buchner began to study the ability of yeast extracts that lacked any living yeast cells to ferment sugar. In a series of experiments at the University of Berlin, he found that the sugar was fermented even when there is no living yeast cell in the mixture. He named the enzyme that brought about the fermentation of sucrose as *zymase*. In 1907, he received the Nobel Prize in chemistry for his biochemical research and discovery of cell-free fermentation.

The first application of cell-free enzymes was the use of rennin isolated from calf or lamb's stomach. Rennin is an aspartic protease which coagulates milk protein and has been used for hundreds of years by the cheese-making industry. Germany prepared the first commercial enzyme in 1914 (Market Research, BBC). This trypsin enzyme was isolated from animals' degraded proteins and used in detergent to degrade proteins. It proved to be powerful as compared to the traditional washing powders used by German house wives which came in small packets.

The first enzyme molecule to be isolated in pure crystalline form was urease, prepared from the jack bean in 1926 by American biochemist J. B. Sumner. The first commercial *Bacillus* protease was marketed in 1959 and became big business when Novozymes in Denmark started to manufacture it and then the major detergent manufacturers started to use it by 1965. Enzymes were used in 1930 in fruit juice manufacturing for clarification and its major usage started in 1960s in starch industry. The use of enzymes results in many benefits such as higher product quality and lower manufacturing cost, and less waste and reduced energy consumption. More traditional chemical treatments produce undesirable side effects and/or waste disposal problems. The degree to which a desired technical effect is achieved by an enzyme can be controlled through various means, such as dose, temperature and time.

## 1.3   INDUSTRIAL ENZYMES: DEVELOPMENT OF BIOINDUSTRIAL SECTOR

The field of industrial enzymes is now undergoing major research and development initiatives, resulting in both the development of number of new products and improvement in the process and performance of several existing products. The global market for industrial enzymes was estimated to touch $ 2.7 billion in 2012. The average annual growth rate (AAGR) of industrial enzymes is about 4–5%, which is accompanied by decrease in prices due to the increased number of manufacturers. The total industrial enzyme market in 2012 was about $ 2.5 billion (Fig. 1.1). The industrial enzyme market is divided into three application segments: technical enzymes, food enzymes and animal feed enzymes.

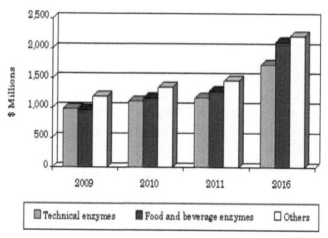

**FIGURE 1.1**   Development of bioindustrial sector, 2009–2016, BCC Research.

The growth of animal feed enzymes is somewhat higher, expected to be close to 4% AAGR, which, in large part, is boosted by increased use of phytase enzyme as an animal feed additive in diets for monogastric animals, for high-temperature feed pelleting processes.

## 1.3.1   GLOBAL AND INDIAN SCENARIO OF INDUSTRIAL ENZYMES

Biotechnology is gaining rapid ground as it offers several advantages over conventional technologies. The global market for industrial enzymes is estimated at $ 3.3 billion in 2010 and expected to reach $ 4.4 billion by 2015. Technical enzymes are valued at just over $ 1 billion in 2010. This sector will increase at a 6.6% compound annual growth rate (CAGR) to reach $ 1.5 billion in 2015. The highest sales of technical enzymes occurred in the leather market, followed by the bioethanol market. The food and beverage enzymes segment is expected to reach about $1.3 billion by 2015, from a value of $ 975 million in 2010, rising at a CAGR of 5.1%. Within the food and beverage enzymes segment, the milk and dairy market had the highest sales, with $ 401.8 million in 2009. The global industrial enzyme market segmentation for various areas of application shows that 41% of the market is for pharmaceuticals followed by detergent and cleaners (17%), food and feed 17% whereas 17% captured by the leather and paper industries and 8% by textile industries (Fig. 1.2).

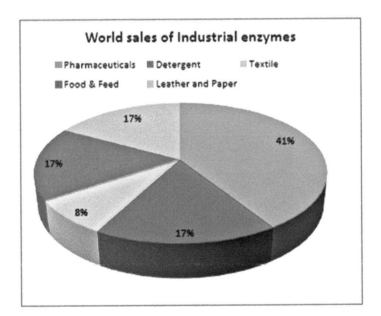

**FIGURE 1.2** Global consumption of industrial enzymes.
*Source:* Binod et al., 2013

Indian biotechnology is now poised to leverage its scientific skills and technical experiences to make a global impact on a strong innovation-led platform. Apart from the technological capabilities, the firm must be able to position its product in pharmaceutical industries, followed by food/feed and textile markets. The market, thus, serves as the link between consumer needs and the pattern of industrial response. The biotech industry in India (Fig. 1.3) accounts for just 2% of global biotech markets. However, it is gaining global visibility because of the investment opportunities. In India, the industrial enzyme consumption is predominantly in the pharmaceutical market (50%), followed by the detergent market (20%). The other important segments are food and feed, textiles, leather, and pulp and paper. In the recent years, enzymes have found numerous applications in the food, pharmaceutical, diagnostic and chemical processing industries.

### Indian scenario of Industrial enzymes demand

■ Pharmaceuticals ■ Detergent ▒ Textile ■ Food & Feed ▒ Leather and Paper

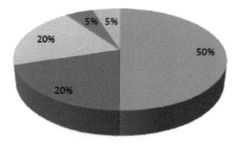

**FIGURE 1.3**   Consumption of industrial enzymes in India.
*Source:* Binod et al., 2013

It is a powerful enabling technology that can transform agriculture and healthcare, use renewable resources to bring greater efficiency into industrial processes, check environmental degradation and deliver a more bio-based economy. With the rise of environment consciousness and regulations, manufacturers have realized that eco-friendly chemical alternatives are way forward, especially if they wish to penetrate untapped and niche areas such as the laundry bar market. Industrial enzyme manufacturers need to realize that consumers are likely to purchase ecological products only if they are cost effective and easy to use.

## 1.4 CLASSIFICATION OF ENZYMES

A more rational system of classification of enzymes divides them into the following six major classes based on the type of reactions they catalyse.

1.  Oxidoreductases: These catalyse oxidation and reduction reactions.
2.  Transferases: These catalyse the transfer of a chemical group from one compound to another.
3.  Hydrolases: These catalyse hydrolysis reaction.
4.  Lyases: These catalyse the breaking of various chemical bonds of molecules other than hydrolysis.
5.  Isomerases: These catalyse the transesterification of one isomer into another.
6.  Ligases: These catalyse the formation of bonds between compounds often using the free energy made available from adenosine triphosphate (ATP) hydrolysis.

### 1.4.1 INDUSTRIALLY IMPORTANT ENZYMES

Microorganisms represent an excellent source of enzymes due to their broad biochemical diversity and their susceptibility to genetic manipulation. Among the broad source of enzymes, that is animals, plants and microbes, enzymes from microorganisms have become the choice of industrial production. Microbial enzymes provide vast diversity of catalytic activities and can be produced more economically. Microbial enzymes have not only generated significant scientific knowledge but have also shown their enormous potential in biotechnology. With the advancement in biotechnology, especially in the area of genetics and protein engineering, a new era of enzyme applications in various industrial processes has begun with major research and development initiatives, resulting not only in the development of new products but also performance of several existing processes. Industrially important enzymes may include:

#### 1.4.1.1 PROTEASES

Proteases are enzymes that act on proteins and mainly involved in hydrolysis of peptides bonds. They play an important role in food, pharmaceutical, leather, textile, detergent, and diagnostics industries and also in waste management.

## 1.4.1.2 AMYLASES

Amylases act on starch, amylose and amylopectin. They break down starch into dextrins and sugars by cleaving the $\alpha$-1,4-glycosidic linkages in the interior of the starch chain. They play a major role in food and beverages, brewing, starch, sugar and alcohol industries.

## 1.4.1.3 LIPASES

Lipases catalyse ester hydrolysis, ester synthesis, transesterification and other reactions. Lipase hydrolyses insoluble fats and fatty acid esters to yield monoglycerides, glycerides, glycerol and free fatty acids. Lipases play major role in food, detergent, bakery, paper and pulp, oleochemicals, cosmetics, textile, leather, diagnostic tool, biodiesel and waste treatment industries. Lipases from fungal and bacterial sources have been purified (Pandey et al., 2010). Lipases from fungal and bacterial sources have been purified (Pandey et al., 2010). Lipases are acidic glycoproteins of molecular weights ranging from 20,000–60,000 Da. The specific activities of the pure proteins vary from 1000–10,000 units per milligram protein. The majority of purified lipases contain 2–15% carbohydrate side chains that are probably not associated with the catalytic activity of enzymes. The autolysis of *Rhizopus arrhizus* lipase gives a low-molecular weight glycopeptide and a carbohydrate free of protein with high lipase activity (Frenking et al., 2011). The multiple forms of lipases apparently produced by some animal tissues and microorganisms may result from self-association of enzyme molecules or variation in the carbohydrate side chains. The low proportion of hydrophobic residues in lipases with strong interaction with hydrophobic substances at an interface is probably caused by hydrophobic patches on their surfaces. Such patches may also be responsible for the self-association behaviour shown by the enzymes in aqueous solutions.

## 1.4.1.4 CELLULASES

Cellulase acts on cellulose molecules by hydrolysing the $\beta$-1, 4-glycosidic linkages. It largely produces cellobiose which can ultimately yield glucose units. Cellulases have wide application in the food and beverages industries.

## 1.4.1.5   PECTINASES

Pectinases act on pectins and their derivatives, and play a major role in the food and beverages industry. Pectinase is a unique enzyme used in fruit drinks for the peeling of citrus fruit.

## 1.4.1.6   XYLANASES

Xylanases cleave chains of $\beta$-1, 4-xylosidic linkages in xylans. They are mainly used in the food and beverages, and baking industries.

## 1.4.1.7   GLUCOSE OXIDASES

Glucose and hexose oxidases catalyse reactions between glucose and oxygen producing gluconolactone and hydrogen peroxide. They are mainly used in food industries.

## 1.4.1.8   CATALASES

Catalase is the enzyme that breaks down hydrogen peroxide to water and molecular oxygen. Catalase effectively removes the residual hydrogen peroxide ensuring that the fabric is peroxide free and mainly used in food industry and also in egg processing with other enzymes.

## 1.4.1.9   GLUCANASES

Glucanase act on $\beta$-1,3 and $\beta$-1,4 bonds in $\beta$-D-glucans. $\beta$-glucanases are of particular interest to the brewing industry, where they act on the glucans that impede clarification of wort and filtration of beer.

## 1.4.1.10   HEMICELLULASES

Hemicellulases act on hemicellulose, a polymer of pentose sugars. They are mainly used in the food and beverages industry to improve the quality of dough, the softness of the crumb and volume. Indeed, the demand for these enzymes is growing more rapidly than ever before, and this demand has become the driving force for research on these enzymes.

## 1.5  LIPASE

With the increase in demand for various enzymes in industries, lipase is considered as the fastest growing important group of biotechnologically relevant enzymes due to its unique characteristics which include substrate specificity, stereospecificity, regioselectivity and ability to catalyse a heterogeneous reaction at the interface of water soluble and water insoluble systems (Treacy et al., 2001). Lipase catalyses range of bioconversion reactions which include hydrolysis, interesterification, esterification, alcoholysis, acidolysis and aminolysis (Jaeger et al., 1994; Quinn et al., 1983). This versatility makes lipases the enzymes of choice for potential applications in the food, detergent, pharmaceutical, leather, textile, cosmetics and paper industries.

Bacterial lipases play a vital role in commercial ventures. Lipases are serine hydrolases and have high stability in organic solvents. The present trend in lipase research is the development of novel and improved lipases through molecular approaches such as directed evolution and exploring natural communities by the metagenomic study.

Lipases (triacylglycerol hydrolases EC 3.1.1.3) belong to a subclass of esterases that catalyse the hydrolysis of triacylglycerols (TAGs) to glycerol and fatty acids (FAs). Lipases are ubiquitous in nature and aid in processes such as digestion, membrane phospholipid metabolisms and inflammatory reactions (Vance and Vance, 2002). Plants, animals and microorganisms produce lipases. Animal lipases are found in several different organs such as the pancreas and digestive tract. Recently, more attention is being paid to lipases produced by fungi. Microbial lipases are relatively stable and are capable of catalysing a variety of reactions; they are of potential importance for diverse industrial applications. There are many reasons for this growing interest in enzyme-mediated reactions compared to chemical processes, including high degree of specificity, mild reaction conditions, decrease in side reactions and simplicity of post-recuperation. Furthermore, enzyme-mediated processes are energy saving and reduce the extent of thermal degradation (Pandey et al., 1999; Osborn and Akoh, 2002). Bacteria produce various families of lipolytic enzymes including true lipases (EC 3.1.1.3), carboxylesterases (EC 3.1.1.3), secretory lipases (EC 3.1.1.3) and a variety of phospholipases namely, phospholipase A1 (EC 3.1.1.32), phospholipase A2 (EC 3.1.1.4), lysophospholipase (EC3.1.1.5), phospholipase C (EC 3.1.4.11) and phospholipase D (EC 3.1.4.4).

The following developments have advanced the knowledge of the structure and function of lipases and esterases in following ways:

1. Increased availability of gene sequences
2. Biochemical characterization of lipases
3. Resolution of numerous crystal structures

### 1.5.1   KINETIC MODEL OF LIPOLYSIS

Lipolysis occurs at the substrate–water interface and therefore cannot be described by the Michaelis–Menten model, which is applicable only for biocatalysis in a homogeneous phase in which the substrate and the enzyme are soluble.

Many different bacterial species produce lipases that hydrolyse esters of glycerol with preferably long-chain FAs. They act at interface generated by a hydrophobic lipid substrate in a hydrophilic aqueous medium. A characteristic property of lipases is interfacial activation. This is a sharp increase in lipase activity when the substrate starts to form an emulsion, thereby presenting to the enzyme an interfacial area and consists of two successive equilibria. In the first equilibrium phase, reversible adsorption of the enzyme to the interface (E↔E*) occurs, while in the second phase, the adsorbed enzyme (E*) binds a single substrate molecule (S), resulting in the formation of a (E*S) complex. This equilibrium is equivalent to the Michaelis–Menten equilibrium for the enzyme–substrate complex. Once the (E*S) complex is formed, the subsequent catalytic steps take place, ending with the release of the products and regeneration of the enzyme in the (E*) form. The synthesis and secretion of lipases by bacteria is influenced by a variety of environmental factors such as ions, carbon sources or presence of non-metabolizable polysaccharides.

### 1.5.2   STRUCTURE OF LIPASE

Lipases from fungal and bacterial sources have been purified (Brockerhoff and Jensen, 1974; Prabhakar et al., 2002). Lipases are acidic glycoproteins of molecular weights ranging from 20,000–60,000 Da. The specific activities of the pure proteins vary from 1000–10,000 units per milligram protein. The majority of purified lipases contain 2–15% carbohydrate side chains that are probably not associated with the catalytic activity of enzymes. The autolysis of R. *arrhizus* lipase gives a low-molecular weight glycopeptide and a carbohydrate free of protein with high lipase activity (Semeriva et al., 1967).

The multiple forms of lipases apparently produced by some animal tissues and microorganisms may result from self-association of enzyme molecules or variation in the carbohydrate side chains. The low proportion of hydrophobic residues in lipases with strong interaction with hydrophobic substances at an interface is probably caused by hydrophobic patches on their surfaces. Such patches may also be responsible for the self-association behaviour shown by the enzymes in aqueous solutions (Fig. 1.4).

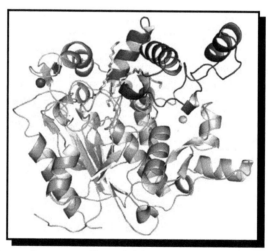

**FIGURE 1.4**   A crystal structure of bacterial thermoalkalophilic lipases from *Geobacillus thermocatenulatus.*

### 1.5.3   TEMPERATURE, pH OPTIMA, STABILITY AND COFACTOR REQUIREMENT

Lipases are reasonably stable in neutral aqueous solutions at room temperature. Solutions of pancreatic and many extracellular microbial lipases loose activity on storage at room temperature above 40°C, but some microbial lipases are more resistant to heat inactivation. The enzyme produced by *Aspergillus niger*, *Rhizopus japonicus* and *Chromobacterium viscosum* are stable in solution at 50°C and the thermotolerant fungus *Humicola lanuginosa* excretes a lipase that is stable at 60°C (Fukumoto et al, 1963; Aisaka and Terada, 1980; Yamaguchi and Mase, 1991; Liu et al., 1973). A strain of *Pseudomonas nitroreducens* gives a lipase which is stable at 70°C, and the *Pseudomonas fluorescens* enzyme, which can be responsible for spoilage of heat-treated dairy products, is only partially irreversibly inactivated at high temperatures (Watanabe et al., 1977; Adams and Brawley, 1981).

Lipases have a broad pH activity profile, showing high activity between pH 5.0 and 9.0 with a maximum between pH 6.0 and 8.0. The extracellular lipases produced by *A. niger*, *C. viscosum* and *R. arrhizus* are particularly active at low pH and an alkaline lipase active at pH 11.0 was isolated from *P. nitroreducens* (Laboureur and Labrousse, 1966). Cofactors are not required for the expression of lipase activity although substances which affect the amount or properties of the interface between the substrate and the aqueous phase have an effect on the reaction rate. Salts can have pronounced effect on the lipase reaction by influencing the ionization of the FA product. In particular, calcium ions often stimulate the reaction by removing, as in calcium soaps, FA anions which are inhibitory.

Lipases display resistance to organic solvents including lipase/esterase from *Pyrobaculum calidifontis*, *Bacillus licheniformis*, *Burkholderia cepacia*, *Arthrobacter nitroguajacolicus* and other *Bacillus* sp. (Sharma et al., 2001). These enzymes show remarkable stability and activity in 50% (v/v) organic solvents such as acetone, methanol, acetonitrile, propanol and dimethyl sulphoxide. The substrate specificity of lipase is often crucial to their application for analytical and industrial purposes. The enzyme can show specificity with respect to either the fatty acyl or alcohol parts of their substrates.

## 1.6   CLASSIFICATION OF BACTERIAL LIPOLYTIC ENZYMES

Bacteria produce different classes of lipolytic enzymes, including carboxylesterases (EC 3.1.1.1), which hydrolyse small ester-containing molecules that are partly soluble in water and true lipases (EC 3.1.1.3) which display maximal activity towards water insoluble long-chain triglycerides, and various types of phospholipase (Arpigny and Jaeger, 1999). Classification of esterases and lipases contribute to a faster identification and an easier characterization of novel bacterial lipolytic enzymes into three different ways:

1.  Important structural features such as residues forming the catalytic site or the presence of disulphide bonds.
2.  Types of secretion mechanism and requirement for lipase-specific foldases (Lif).
3.  Potential relationship to the other enzyme families.

Based on the comparison of their amino acid sequences and some fundamental biological properties, bacterial esterases and lipases are classified into eight different families with the largest being further divided into six subfamilies.

## 1.6.1   FAMILY I

### 1.6.1.1   TRUE LIPASE (PSEUDOMONAS LIPASES)

True lipases (EC 3.1.1.3) hydrolyse water insoluble long-chain triglycerides into glycerols and FAs. Bacterial true lipases were formerly included in the so called *Pseudomonas* groups 1, 2 and 3 because *Pseudomonas* lipases were probably the first to be studied and had a preponderant role in the industry. Some *Pseudomonas* species that produce important lipases have recently been renamed *Burkholderia* (Yabuuchi et al., 1992). True lipases or *Pseudomonas* lipases are divided into three subfamilies I.1, I.2 and I.3.

#### 1.6.1.1.1   Subfamily I.1

The molecular mass of these enzyme are in the range 30–32 kDa. *Pseudomonas aeruginosa* lipases provided the first structure in the lipase subfamily I.1. Lipases from *Vibrio cholerae, Acinetobacter calcoaceticus, Proteus vulgaris* and *Pseudomonas wisconsinensis* have molecular mass of 30–32 kDa and displayed a higher sequence similarity to the *P. aeruginosa* lipase.

### *1.6.1.1.2   Subfamily I.2*

The molecular mass of enzymes from I.2 subfamily are characterized by a slightly larger size, 33 kDa, owing to an insertion in the amino acid sequence forming an antiparallel double β-strand on the surface of the molecule. *Burkholderia glumae* (Noble et al., 1993) was the first bacterial lipase with known three-dimensional structure and belongs to subfamily I.2. Lipases from *C. viscosum*, *Burkholderia cepacia* and *Pseudomonas luteola* show high similarity to the *Burkholderia* enzymes and come under subfamily I.2 (Lang et al., 1996; Mala and Takeuchi, 2008).

The expression in an active form of lipases belonging to subfamilies I.1 and I.2 depends on a chaperone protein named Lif. Both subfamilies I.1 and I.2 share important structural features such as cysteine residue from disulphide bridge and aspartic residue involved in the $Ca^{2+}$ binding site. The residues involved in the formation of both the $Ca^{2+}$ binding site and the disulphide bridge are located in the vicinity of the catalytic histidine (His) and aspartic acid (Asp) residues; they are believed to be important in the stabilization of the active centre of these enzymes (Bhardwaj et al., 2006).

### *1.6.1.1.3   Subfamily I.3*

Subfamily I.3 contains enzymes from at least two distinct species: *P. fluo-rescens* and *Serratia marcescens*. These lipases have in common a higher molecular mass than lipase from subfamilies I.1 and I.2. *Pseudomonas fluorescens* lipase has molecular mass 50 kDa, and *S. marcescens* lipase has molecular mass 65 kDa. The secretion of these enzymes occurs in one step through a three-component ATP-binding-cassette transporter system (Duong et al., 1994; Li et al., 1995).

### *1.6.1.2   FAMILY II (GDSL FAMILY)*

The enzymes grouped in family II do not exhibit the conventional penta-peptide Gly-Xaa-Ser-Xaa-Gly but rather display a Gly-Asp-Ser-(Leu) (GDSL) motif containing the active site serine residue, and this residue lies much closer to the N-terminus than in other lipolytic enzymes. Crystal structure of the catalytic centre of *Streptococcus scabies* esterase has a particular architecture, in that it forms a catalytic dyad instead of a triad (Wei et al., 1995). The acidic side chain of the active site His residue is replaced by the backbone carbonyl of the residue located three positions upstream of the His itself namely Trp-315. Platelet-activating factor acetyl-hydrolase (α1PAF-AH) from bovine brain and *Aeromonas hydrophila* esterases belong to family II. Another interesting feature of the GDSL esterases from *P. aeruginosa* and *Salmonella typhimurium* is the domain that encompasses approximately one-third of their entire sequences and is similar to that of a newly identified family of autotransporting bacterial virulence factors (Loveless and Saier, 1997; Henderson et al., 1998).

### 1.6.1.3   FAMILY III

This family of lipase was identified primarily by Cruz et al. (1994) and mentioned by Wei et al. (1998) who solved the three-dimensional struc-ture of *Streptococcus exfoliates* (M11) lipase. This enzyme displays the canonical fold of α/β-hydrolases and contains a typical catalytic triad. It also shows approximately 20% amino acid sequence identity with the intracellular and plasma isoforms of the human PAF-AH. These PAF-AHs are monomer proteins, in contrast with the heterotrimeric PAF-AH from bovine brain. Their active site Asp residue, identified primarily by site-directed mutagenesis, was shown to be located in the sequence at a position nonequivalent to that found in the *S. exfoliates* enzyme again underlining the great functional versatility of the α/β-hydrolase scaffold.

### 1.6.1.4   FAMILY IV [HORMONE-SENSITIVE LIPASE (HSL) FAMILY]

A number of bacterial enzymes of this family display a striking amino acid sequence similarity to the mammalian hormone-sensitive lipase (HSL) family (Hemila et al., 1994). These enzymes show relatively high activity at low temperature (less than 15°C) retained by HSL and the lipase from *Moraxella* sp. (Feller et al., 1991). The mammalian HSL is derived from a catalytic domain, homologous with the bacterial enzymes, merged with an additional N-terminal domain and a regulatory module inserted in the central part of the sequence.

### 1.6.1.5   FAMILY V

Enzymes grouped in family V originate from mesophilic bacteria (*Pseudomonas oleovorans, Haemophilus influenzae* and *Acetobacter pasteurianus*) as well as cold-adapted (*Moraxella* sp., *Psy. immobilis*) or heat-adapted (*Sulfolobus acidocaldarius*) organisms. They share significant amino acid sequence similarity (20–25%) to various bacterial non-lipolytic enzymes, namely epoxide hydrolases, dehalogenases and haloperoxidase, which also possess the typical α/β-hydrolase fold and a catalytic triad (Verschueren et al., 1993).

### 1.6.1.6   FAMILY VI

The enzymes belonging to family VI with a molecular mass in the range 23–26 kDa are known to be the smallest esterases. *Pseudomonas fluorescens* carboxylesterase is a member of this family. This carboxylesterase hydrolyses small substrates with a broad specificity and displays no activity towards long-chain triglycerides (Hong et al., 1991). The enzymes in family VI display approx. 40% sequence similarity to eukaryotic lysophospholipases ($Ca^{2+}$ independent phospholipase A2).

### 1.6.1.7   FAMILY VII

Enzymes grouped in family VII are bacterial esterases with molecular mass 55 kDa and share significant amino acid sequence homology (30% identity, 40% similarity) with eukaryotic acetylcholine esterase and intestine liver carboxylesterases. The esterases from *Arthrobacter oxydans*, *Bacillus subtilis* and *Streptomyces coelicolor* belong to the family VII.

### 1.6.1.8   FAMILY VIII

Three enzymes forming this family are approximately 380 residues long and show a striking similarity to several class C β-lactamases. A stretch of 150 residues is notably 45% similar to an *Enterobacter cloacae ampC* gene product. This feature suggests that the esterases in family VIII possess an active site more reminiscent of that found in class C β-lactamases, which involves a Ser-Xaa-Xaa-Lys motif conserved in the N-terminal part of both enzyme categories (Galleni et al., 1988).

## 1.6.2   LIPASE FROM GRAM-POSITIVE BACTERIA

In *Bacillus* lipase, an alanine residue replaces the first glycine in the conserved pentapeptide: Ala-Xaa-Ser-Xaa-Gly. However, the lipase from the two mesophilic strains *B. subtilis* and *B. pumilus* stand apart because they are the smallest true lipases known (approx 22 kDa) and share very little sequence similarity of approximately 15% with the other *Bacillus* and *Staphylococcus* lipases. *Bacillus stearothermophilus* and *B. thermocatenulatus* produce lipase with similar properties having a molecular mass of approximately 45 kDa and display maximal activity at pH 9.0 and 65°C temperature (Kim et al., 1998). *Staphylococcal* lipases are larger enzymes (approx 75 kDa) that are secreted as precursors and cleaved in the extracellular medium by a specific protease, yielding a mature protein of approximately 400 residues. Lipase from *Staphylococcus hyicus* (Table 1.1) displays a remarkable phospholipase activity which is unique among true lipases (Van-Oort et al., 1989).

**TABLE 1.1**   Lipases from Various Microorganisms.

| | |
|---|---|
| Bacteria (Gram-positive) | *Bacillus* species: *B. subtilis, B. thermoleovorans, B. thermocatenulatus, B. coagulans, Lactobacillus plantarum* |
| | *Staphylococcus* species: *S. haemolyticus, S. aureus, S. warneri, S. xylosus* |
| | *Enterococcus faecalis* |
| Bacteria (Gram-negative) | *Pseudomonas* species: *P. aeruginosa, P. fluorescens, P. fragi* |
| | *Penicillium* species: *P. cyclopium, P. simplicissimum* |
| Fungi | *Aspergillus niger, Aspergillus oryzae, Rhizopus* sp., *Rhizomucor miehei, Geotrichum candidum, Pichia burtonii* |
| Yeast | *Candida cylidracae* |

### 1.6.3   OTHER LIPASES

The lipases from *Propionibacterium acnes* (339 residues) and *Streptomyces cinnamoneus* (275 residues) show significant similarity to each other (39% identity, 50% similarity), and 50% similarity to the lipase from *B. subtilis* and from subfamily I.2. There is no similarity found between *Cinnamoneus* lipase and *Streptomyces* lipases known so far (Miskin et al., 1997).

Table 1.1 shows the lipases from different microorganisms.

## REFERENCES

Adams, D. M.; Brawley, T. G. Factors Influencing the Activity of a Heat- Resistant Lipase of *Pseudomonas. J. Food Sci.* **1981,** *46,* 677–680

Aisaka, K.; Terada, O. Purification and Properties of Lipase from *Rhizopus japonicas. Agric. Biol. Chem.* **1980,** *44,* 799–805.

Arpigny, J. L.; Jaeger, K. E. Bacterial Lipolytic Enzymes: Classification and Properties. *Biochem. J.* **1999,** *343,* 177–183.

Bhardwaj, N., et al. Structural Bioinformatics Prediction of Membrane-Binding Proteins. *J. Mol. Biol.* **2006,** *359,* 486–495.

Binod, P., et al. Industrial Enzymes-Present Status and Future Perspectives for India. *J. Digest.* **2013,** *23,* 74–84.

Brady, S. F., Chao, C. J.; Clardy, J. New Natural Product Families from an Environmental DNA (eDNA) Gene Cluster. *Am. Chem. Soc.* **2002,** *124,* 9968–9969.

Brockerhoff, H.; Jensen, R. *The Enzymes*. Academic Press: New York, 1974.

Cruz, H., et al. Sequence of the *Streptomyces albus* G Lipase-Encoding Gene Reveals the Presence of a Prokaryotic Lipase Family. *Gene.* **1994,** *144,* 141–142.

Duong, F., et al. The Pseudomonas Fluorescens Lipase has a C-Terminal Secretion Signal and is Secreted by a Three-Component Bacterial ABC-Exporter System. *Mol Microbiol.* **1994,** *11,* 1117–1126.

Feller, G., et al. Cloning and Expression in *Escherichia coli* of Three Lipase-Encoding Genes from the Psychrotrophic Antarctic Strain *Moraxella* TA144. *Gene.* **1991,** *102,* 111–115.

Frenking, S. G., et al. Effects of Enzyme Replacement Therapy on Growth in Patients with Mucopolysaccharidosis Type II. *J. Inherit. Metab. Dis.* **2011,** *34,* 203–208.

Frost and Sullivan. Research and Markets: European Tactical Communication Market Assessment: in Spite of the Economic Slump the European Tactical Communication Market is Set to Witness Growth. *Business wire*, Dublin, 2010.

Fukumoto, J.; Iwai, M.; Tenjisaka, Y. Purification and Crystallization of a Lipase Secreted by *Aspergillus niger. J. Gen. Appl. Microbiol.* **1963,** *9,* 353–361.

Garrity, G. M.; Bell, J. A.; Lilburn, T. Order II. *Acidithiobacillales* ord. nov. In *Bergey's Manual of Systematic Bacteriology*; 2nd ed., vol 2 (*The Proteobacteria*), part B (*The Gammaproteobacteria*); Brenner, D. J., Krieg, N. R., Staley, J. T., Garrity, G. M. Eds.; Springer: New York, 2005; p 60.

Hemila, H.; Koivula, T. T.; Palva, I. Hormone-Sensitive Lipase is Closely Related to Several Bacterial Proteins, and Distantly Related to Acetylcholinesterase and Lipoprotein Lipase: Identification of a Superfamily of Esterases and Lipases. *Biochim. Biophys. Acta.* **1994,** *1210,* 249–253.

Henderson, I. R.; Navarro-Garcia, F.; Nataro, J. P. The Great Escape: Structure and Function of the Autotransporter Proteins. *Trends Microbiol.* **1998,** *6,* 370–378.

Jaeger, K.E., et al. Bacterial Lipases. *FEMS Microbiol. Rev.* **1994,** *151,* 29–63.

Laboureur, P.; Labrousse, M. Lipase of Rhizopus Aerhizus. Obtaining, Purification and Properties of the Lipase of *Rhizopus arrhizus* Var. Delemar. *Bull. Soc. Chim. Biol.* **1966,** *48,* 747–769.

Lang, D., et al. Crystal Structure of a Bacterial Lipase from Chromobacterium Viscosum ATCC 6918 Refined at 1.6 Å Resolution. *J. Mol. Biol.* **1996,** *259,* 704–717.

Li, X., et al. Gene Cloning, Sequence Analysis, Purification, and Secretion by Escherichia Coli of an Extracellular Lipase Gene. *Appl. Environ. Microbiol.* **1995,** *61,* 2674–2680.

Liu, W. H.; Beppu, T.; Arima, K. Purification and General Properties of the Lipase of Thermophilic Fungus *Humicola lanuginosus* S-38. *Agric. Biol. Chem.* **1973,** *37,* 157–163.

Loveless, B. J.; Saier, M. H. A Novel Family of Channel-Forming, Autotransporting, Bacterial Virulence Factors. *Mol. Membr. Biol.* **1997,** *14,* 113–123.

Mala, J. G.; Takeuchi, S. Understanding Structural Features of Microbial Lipase-An Overview. *Anal. Chem. Insights.* **2008,** *3,* 9–19.

Miskin, J. E., et al. *Propionibacterium* Acnes, a Resident of Lipid-Rich Human Skin, Produces a 33 kDa Extracellular Lipase Encoded by gehA. *Microbiology.* **1997,** *143,* 1745–1755.

Noble, M. E., et al. The Crystal Structure of Triacylglycerol Lipase from *Pseudomonas glumae* Reveals a Partially Redundant Catalytic Aspartate. *FEBS Lett.* **1993,** *33,* 123–128.

Osborn, H. T.; Akoh, C. C. Influence of Molecular Environment on Lipid Oxidation of Structured Lipid-Based Model Emulsions. *Compr. Rev. Food Sci. Food Saf.* **2002,** *3,* 93–103.

Pandey G, et al. (2010) An Integrative Multi-Network and Multi-Classifier Approach to Predict Genetic Interactions. *PLoS Comput. Biol.* **2010,** *6*(9)e1000928. Published online 2010 Sep 9. doi: 10.1371/journal.pcbi.1000928.

Pandey, A., et al. Advanced Strategies for Improving Industrial Enzymes. *Chem. Ind. Sci. Ind. Res.* **2013,** *72,* 271–286.

Pandey, A., et al. The Realm of Microbial Lipases in Biotechnology. *Appl. Biochem. Biotechnol.* **1999,** *29,* 119–131.

Prabhakar, T.; Kumar, A. K.; Ellaiah, P. The Effect of Cultural Conditions on the Production of Lipase by Fungi. *J. Sci. Ind. Res.* **2002,** *61,* 123–127.

Quinn, D.; Shirai, K.; Jackson, R. L. Lipoprotein Lipase: Mechanism of Action and Role in Lipoprotein Metabolism. *Prog. Lipid Res.* **1983,** *22,* 35–78.

Riesenfeld, C. S.; Schloss, P. D.; Handelsman, J. Metagenomics: Genomic Analysis of Microbial Communities. *Annu. Rev. Genet.* **2004,** *38,* 525–552.

Schloss, P. D.; Handelsman, J. Biotechnological Prospects from Metagenomics. *Curr. Opin. Biotechnol.* **2003,** *14,* 303–310.

Semeriva, M.; Benzonana, G.; Desnuelle. P. *Rhizopus arrhizus* Lipase I Positional Specificity. *Bull. Soc. Chim. Biol.* **1967,** *49,* 71–79.

Sharma, R.; Chisti, Y.; Banerjee, U. C. Production, Purification, Characterization and Applications of Lipases. *Biotechnol. Adv.* **2001,** *19,* 627–662.

Treacy, J., et al. Evaluation of Amylase and Lipase in the Diagnosis of Acute Pancreatitis. *ANZ J. Surg.* **2001,** *71,* 577–582.

Vance, D. E.; Vance, J. E. *Biochemistry of Lipids, Lipoprotein and Membranes,* Vol. 36, 4th Ed..; Elsevier Science, 2002, pp 400–612.

Van-Oort, M. G., et al. Purification and Substrate Specificity of *Staphylococcus hyicus* Lipase, *Biochemistry.* **1989,** *28,* 9278–9285.

Verschueren, K. H., et al. Crystallographic Analysis of the Catalytic Mechanism of Haloalkane Dehalogenase. *Nature* (London). **1993,** *363,* 693–698.

Watanabe, N., et al. Isolation and Identification of Alkaline Lipase Producing Microorganisms, Cultural Conditions and Some Properties of Crude Enzymes. *Agric. Biol. Chem.* **1977**, *41*, 1353–1358.

Wei, Y., et al. A Novel Variant of the Catalytic Triad in the Streptomyces Scabies Esterase. *Nat. Struct. Biol.* **1995**, *2*, 218–223.

Wei N, et al. The COP9 Complex is Conserved between Plants and Mammals And is Related to the 26S Proteasome Regulatory Complex. *Curr. Biol.* **1998**, *8*(16), 919–922.

Yabuuchi, E., et al. Proposal of *Burkholderia gen.* nov. and Transfer of Seven Species of the Genus Pseudomonas Homology Group II to the New Genus, with the Type Species *Burkholderia cepacia* (Palleroni and Holmes 1981) comb, nov. *Microbiol Immunol.* **1992**, *36*, 1251–1275.

Yamaguchi, S.; Mase, T. Purification and Characterization of Mono and Diacylglycerol Lipase Isolated from *Penicillium camembertii* U-150. *Appl. Microbiol. Biotechnol.* **1991**, *34*, 720–725.

# CHAPTER 2

# APPLICATION OF LIPASES

## CONTENTS

Microbial lipases constitute an important group of biotechnologically valuable enzymes, mainly because of the versatility of their applied properties and ease of mass production. Microbial lipases are widely diversified in their enzymatic properties and substrate specificity, which make them very attractive option for industrial applications.

Lipases are valued biocatalysts because they act under mild conditions, are highly stable in organic solvents, show broad substrate specificity and usually show high regio- and/or stereoselectivity in catalysis. Lipase catalysed processes offer cost-effectiveness too in comparison with traditional downstream processing. Lipases remain active in organic solvents in their industrial applicability. Lipases form an integral part of the industries ranging from food, dairy, pharmaceuticals, agrochemical, detergents, polymeric materials, oleochemicals, tea industries, cosmetics, leather and in several bioremediation processes.

The cost of lipase production is a major obstacle in industry for its successful application. Lipase yields have been improved by screening for hyperproducing strains and/or by optimization of the fermentation medium. Strain improvements by either conventional mutagenesis or recombinant deoxyribonucleic acid (DNA) technology have been useful in enhancing the production of lipase. Increase in the yield of lipase is important for developing therapeutic agents against diseases such as familial lipoprotein lipase and acid lipase diseases and lysosomal acid lipase (LAL) deficiency.

The biodiversity represents an invaluable resource for biotechnological innovations and plays an important role in the search for the improved strains of microorganisms used in the industry. A recent trend has involved contributing industrial reactions with enzymes reaped from exotic microorganisms that inhabit hot waters, freezing arctic water, saline water or extremely acidic or alkaline habitats. Exploitation of biodiversity to provide microorganisms that produce lipase well suited for their diverse applications is considered to be one of the most promising future alternatives.

Lipase from *Chromobacter, Alcaligenes, Arthrobacter, Bacillus, Burkholderia, Chromobacterium, Pseudomonas* and *Staphylococcus* are widely used for a variety of biotechnological applications.

## 2.1  DETERGENT INDUSTRY

The most commercially important field of application for hydrolytic lipases is their incorporation in detergents, which are used mainly in household and industrial laundry and household dishwashers. Laundry detergents are becoming more and more popular because of their increasing use in washing machines, where they impart softness, resiliency to fabrics, and are antistatic, dispersible in water and mild to eyes and skins. Lipase is used in detergent formulations to remove fat-containing stains such as those resulting from frying fats, salad oils, butter, fat-based sauces, soups, human sebum or certain cosmetics.

## 2.2  FOOD PROCESSING, FLAVOUR DEVELOPMENT AND IMPROVING QUALITY

The lipases play an important role in the fermentative steps of sausage manufacture and to determine changes in long-chain fatty acid liberated during ripening. Earlier, lipases of different microbial origin had been used for refining rice flavour, modifying soybean milk and for improving the aroma and accelerating the fermentation of apple wine (Aravindan et al., 2007). Lipases are extensively used in the dairy industry for hydrolysis of milk fat. The dairy industry uses lipases to modify the fatty acid chain lengths and to enhance the flavours of various cheeses. Current applications also include the acceleration of cheese ripening and the lipolysis of butter, fat and cream (Mamura and Kitaura, 2000). Lipases hydrolyse lipids to raise the level of flavour moieties.

## 2.3  BAKERY INDUSTRY

Recent findings suggest that phospholipases can be used to substitute or supplement traditional emulsifiers since the enzymes degrade polar wheat lipids to produce emulsifying lipids in situ (Kirk et al., 2002). Texture and softness could be improved by lipase catalysation. Lipase is used to increase the specific volume of breads. Yeast with bacterial lipase gene *LIP A* resulted in higher productivity of enzyme and found use in bread making as a technological additive (Keskin et al., 2005). Increased butter flavour for baked goods was generated by hydrolysis of butter fat with suitable lipases (Uhling, 1998).

## 2.4  FAT AND OLEOCHEMICAL INDUSTRY

The lipase catalysed transesterification in organic solvents is an emerging industrial application such as production of cocoa butter equivalent, human milk fat substitute "Betapol", pharmaceutically important poly-unsaturated fatty acids (PUFA), rich/low-calorie lipids, "designer fats or structured lipid" and production of biodiesel from vegetable oils. Conversion of palm oil into cocoa butter fat substitute can be achieved by inter-esterification and is now a commercial process (Nakajima et al., 2000). The adsorption of lecithin, together with lipase onto a carrier, was effective for conducting the interesterifying reaction efficiently for edible oils and fats. The saving of energy and minimization of thermal degradation are probably the major attractions in replacing the current chemical technologies with biological ones. The scope for the application of lipases in the oleochemical industry is enormous as it saves energy and minimizes thermal degradation during hydrolysis, glycerolysis and alcoholysis (Ghosh et al., 1996).

## 2.5   TEXTILE INDUSTRY

Lipases are used in the textile industry to assist in the removal of size lubricants, in order to provide a fabric with greater absorbency for improved levelness in dyeing. Its use also reduces the frequency of streaks and cracks in the denim abrasion systems. In the textile industry, polyester has certain key advantages including high strength, soft hand, stretch resistance, stain resistance, machine washability, wrinkle resistance and abrasion resistance. Synthetic fibers have been enzymatically modified for the use in the production of yarns, fabrics, textiles, rugs and other consumer items.

## 2.6   COSMETIC INDUSTRY

Lipases have potential application in cosmetics because it demonstrates activities in surfactants and in aroma production (Metzger and Bornscheuer, 2006). Retinoids are of great commercial potential in cosmetics and pharmaceuticals such as skin care products. Water-soluble retinol derivatives were prepared by catalytic reaction of immobilized lipase. Lipases have been used in hair waving preparation; they have also been used as a component of topical nonobese creams or as oral administration (Maugard et al., 2002; Smythe, 1951).

## 2.7  PULP AND PAPER INDUSTRY

The pulp and paper industry processes huge quantities of lignocellulosic biomass every year. The technology for pulp manufacture is highly diverse and numerous opportunities exist for the application of microbial enzymes. The enzymatic pitch control method using lipase has been used in large-scale paper-making process as a routine operation since early 1990s (Bajpai, 1999). Lipase for wastepaper deinking can increase the mashing rate of pulp, increase whiteness and intensity, decrease chemical usage, prolong equipment life, reduced pollution levels of waste water, energy and time efficiency and reduced composite cost (Fig. 2.1).

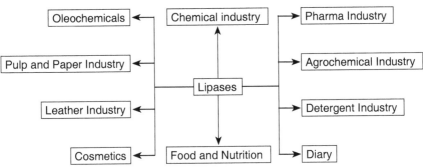

**FIGURE 2.1**  Applications of bacterial lipases.

## 2.8  OIL BIODEGRADATION

Biodegradation of petroleum hydrocarbons in cold environments, including Alpine soils, is a result of indigenous cold-adapted microorganisms' ability to degrade these contaminants. Bacterial monocultures isolated from lubricant-contaminated wastewater of an electric power station have showed positive response in bio-augmented clean-up of waste water contaminated with hydrocarbons and organic polymers using hydrolytic enzymes.

## 2.9  PRODUCTION OF BIODEGRADABLE POLYMERS

Lipases have become one of the most important enzymes used in organic syntheses. Lipases can be used as biocatalysts in the production of useful biodegradable compounds. The ability of lipase to catalyse ester synthesis and transesterification reactions in organic solvent systems has opened up the possibility of enzyme-catalysed production of biodegradable polyesters. Aromatic polyesters can be synthesized by lipase biocatalysis (Linko et al., 1998).

## 2.10  DIAGNOSTIC TOOL

Lipases are important drug targets or marker enzymes in the medical sector. They can be used as diagnostic tools and their presence or increased levels can indicate certain infections or diseases. Lipases are used in the enzymatic determination of serum triglycerides to generate glycerol which is subsequently determined by enzyme-linked colorimetric reactions. The level of lipases in blood serum can be used as a diagnostic tool for detecting conditions such as acute pancreatitis and pancreatic injury (Munoz and Katerndahl, 2000). Acute pancreatitis usually occurs as a result of alcohol abuse or bile duct obstruction. Although serum trypsin level ultrasonography, computed tomography and endoscopic retrograde cholangiopancreatography are the most accurate laboratory indicators for pancreatitis but serum amylase and lipase levels are still used to confirm the diagnosis of acute pancreatitis (Lott and Lu, 1991).

## 2.11  DEGREASING OF LEATHER

Degreasing is an essential stage in the processing of fatty raw materials such as small animal skins and hides from intensively fed cattle. Lipases represent a more environmentally sound method of removing fat. For sheep skins, which contain up to 40% fat, the use of solvents is very common and these can also be replaced with lipases. Lipase can remove fats and grease from skins and hides, particularly those with a moderate fat content. These also hydrolyse triglyceride (the main form of fat stored in animal skins) to glycerol and free fatty acids. Lipases also improve the production of hydrophobic leather; makers of leather for car upholstery have commented that "fogging" is reduced. Lipases offer the tanner two advantages over solvents or surfactants, which are: fat dispersion and production of waterproof and low-fogging leathers.

## 2.12   WASTE/EFFLUENT/SEWAGE TREATMENT

Lipases are utilized in activated sludge and other aerobic waste processes where thin layers of fats must be continuously removed from the surface of aerated tanks to permit oxygen transport. This skimmed fat-rich liquid is digested with lipases such as that from *Candida rugosa* (Bailey and Ollis, 1986). Effective breakdown of solids and the clearing and prevention of fat blockage or filming in waste systems are important in many industrial operations. Examples include:

1.   Degradation of organic debris using a commercial mixture of lipase, cellulase, protease, amylase, inorganic nutrients, wheat bran and so forth is employed for this purpose.
2.   Sewage treatment, cleaning of holding tanks, septic tanks, grease traps, and so forth. Effluent treatment is also necessary in industrial processing units, such as abattoirs, the food processing industry, the leather industry, the poultry waste processing (Godfrey and Reichelt, 1983). Bacterial lipases are involved in such environmental problems solutions by breaking fats in domestic sewage and anaerobic digesters.

## 2.13   PRODUCTION OF BIODIESEL

The limited resources of fossil fuels, increasing prices of crude oil, and environmental concerns have been the diverse reasons for exploring the use of vegetable oils as alternative fuels. The biodiesel fuel from vegetable oil does not produce sulphur oxide and minimize the soot particulate one third times in comparison with the existing one from petroleum. Immobilized *Pseudomonas cepacia* lipase was used for the transesterification of soybean oil with methanol and ethanol (Shah et al., 2004). Novozymes also have been used to catalyse the transesterification of crude soybean oils for biodiesel production in a solvent-free medium (Du et al., 2004). Fatty acids esters were produced from two Nigerian lauric oils, that is, palm kernel oil and coconut oil, by transesterification of the oils with different alcohols using PS30 lipase as a catalyst.

## 2.14   TEA PROCESSING

The quality of black tea is dependent to a great extent on dehydration, mechanical breaking and enzymatic fermentation to which tea shoots are subjected. During the manufacture of black tea, enzymatic breakdown of membrane lipids initiate the formation of volatile products with characteristic flavour properties emphasizing the importance of lipid in flavour development.

## 2.15   BIOSENSORS

A biosensor based on the enzyme-catalysed dissolution of biodegradable polymer films has been developed. The polymer enzyme system, poly(trimethylene) succinate, was investigated for use in the sensor, which is degraded by a lipase. Potential fields of application of such a sensor system are the detection of enzyme concentrations and the construction of disposable enzyme-based immune sensors, which employ the polymer degrading enzyme as an enzyme label (Sumner et al., 2001). Lipases may be immobilized onto pH/oxygen electrodes in combination with glucose oxidase, and these function as lipid biosensors and may be used in triglycerides and blood cholesterol determinations (Karube et al., 1990).

## 2.16   ENVIRONMENTAL MANAGEMENT

The use of lipases in bioremediation processes is a new aspect in lipase biotechnology. Lipase-producing strains play a key role in the enzymological remediation of polluted soils. Cold-adapted lipases have great potential in the field of wastewater treatment, bioremediation in fat-contaminated cold environment and active compounds synthesis in cold condition while in temperate regions, large seasonal variations in temperature reduce the efficiency of microorganisms in degrading pollutants such as oil and lipids. The enzymes active at low and moderate temperature may also be ideal for the bioremediation process (Lin et al., 2012).

## 2.17   MAJOR OBSTACLES AND FUTURE PROSPECT OF MICROBIAL LIPASES

Lipases are the most versatile industrial enzymes and are known to bring about a range of bioconversion reactions. Lipases also serve as biocatalysts for alcoholysis, acidolysis, esterification and aminolysis. Lipases differ in their properties such as substrate specificity, active site and catalytic mechanisms. Their exquisite specificities provide a basis for their numerous physiological and commercial applications.

Despite the extensive research on several aspects of lipase from ancient times, there is tremendous scope of improving their properties to suit projected applications. The future lines of development would include:

1.   Designing of appropriate screening conditions for the isolation of lipase-producing microbes
2.   Development of economical fermentation method(s)
3.   Exploration of lipases suitable for application in industries such as detergent, leather, textile, oleochemical, food, pharmaceuticals and so forth
4.   Determination the physiological role of lipase isoenzymes in the growth and development of lipase-producing bacteria.

## REFERENCES

Aravindan, R.; Anbumathi, P.; Viruthagiri, T. Lipase Applications in Food Industry. *Indian J. Biotechnol.* **2007,** *7,* 141–158.

Bailey, J. E.; Ollis, D. F. Applied Enzyme Catalysis. In *Biochemical Engineering Fundamentals,* 2nd ed.; McGraw-Hill: New York, 1986, pp 157–227.

Bajpai, P. Application of Enzymes in the Pulp and Paper Industry. *Biotechnol. Prog.* **1999,** *15,* 147–157.

Du, W., et al. Novozym 435—Catalysed Transesterification of Crude Soya Bean Oils for Biodiesel Production in a Solvent-Free Medium. *Biotechnol. Appl. Biochem.* **2004,** *40,* 187–190.

Ghosh, P. K., et al. Microbial Lipases: Production and Applications. *Sci. Prog.* **1996,** *79,* 119–157.

Godfrey, T.; Reichelt, J., Eds. Industrial Applications. In: *Industrial Enzymology-Applications of Enzymes in Industry.* The Nature Press: London, 1983; pp 170–465.

Karube, I.; Sode, K.; Tamiya, E. Microbiosensors. *J. Biotechnol.* **1990,** *15,* 267–282.

Keskin, O.; Ma, B.; Nussinov R. Hot Regions in Protein–Protein Interactions: the Organization and Contribution of Structurally Conserved Hot Spot Residues. *J. Mol. Biol.* **2005,** *345,* 1281–1294.

Kirk, O.; Borchert, T. V.; Fuglsang, C. C. Industrial Enzyme Applications. *Curr. Opin. Biotechnol.* **2002,** *13,* 345–351.

Lin, J. F., et al. Bacterial Diversity of Lipase-Producing Strains in Different Soils in Southwest of China and Characteristics of Lipase. *Afr. J. Microbiol. Res.* **2012,** *6,* 3797–3806.

Linko, Y. Y., et al. Biodegradable Products by Lipase Biocatalysis. *J. Biotechnol.* **1998,** *66,* 41–50.

Lott, J. A.; Lu, C. J. Lipase Isoforms and Amylase Isoenzymes—Assays and Application in the Diagnosis of Acute Pancreatitis. *Clin. Chem.* **1991,** *37,* 361–368.

Mamura, S.; Kitaura, S. Purification and Characterization of Monoacylglycerol Lipase from the Moderately Thermophilic *Bacillus* Species H-2575. *J. Biochem.* **2000,** *127,* 419–425.

Maugard, T.; Rejasse, B.; Legoy, M. D. Synthesis of Water-Soluble Retinol Derivatives by Enzymatic Method. *Biotechnol. Prog.* **2002,** *18,* 424–428.

Metzger, J. O.; Bornscheuer, U. Lipids as Renewable Resources: Current State of Chemical and Biotechnological Conversion and Diversification. *Appl. Microbiol. Biotechnol.* **2006,** *71,* 13–22.

Munoz, A.; Katerndahl, D. A. Diagnosis and Management of Acute Pancreatitis. *Am. Fam. Physician.* **2000,** *62,* 164–174.

Nakajima, M.; Snape, J.; Khare, S. K. Applications of Enzymes and Membrane Technology in Fat and Oil Processing. In *Method in Non-Aqueous Enzymology;* Gupta, M. N., Ed.; Birkhauser Verlag: Basel, 2000; pp 52–69.

Shah, S.; Sharma, S.; Gupta, M. N. Biodiesel Preparation by Lipase Catalysed Transesterification of Jatropha Oil. *Energy Fuels.* **2004,** *18,* 154–159.

Smythe, C. V. Microbiological Production of Enzymes and Their Industrial Application. *Econ. Bot.* **1951,** *5,* 126–144.

Sumner, C., et al. Biosensor Based on Enzyme-Catalysed Degradation of Thin Polymer Films. *Biosens. Bioelectron.* **2001,** *16,* 709–714.

Uhling, H. *Industrial Enzymes and Their Applications*, 2nd ed.; John Wiley & Sons: New York, 1998; pp 332–334.

# CHAPTER 3

# METAGENOMICS AND UNCULTURABLE BACTERIA

## CONTENTS

## 3.1    MICROBIAL COMMUNITIES

According to Ferrer et al. (2009) microbes are the most numerous, diverse and dynamic organisms. They constitute the major reservoir of genetic diversity on Earth. Microbes are ubiquitous in every habitat such as water, soil, air, acidic hot spring, glacial ice, highly polluted environment and the most extreme conditions such as volcano vents, Antarctic environments and deep sea hydrothermal vents to rocks in deep boreholes beneath the Earth's surface. Courtois et al. (2003) and Pace et al. (1985) reported that total number of bacterial and archaeal cells on earth has been estimated to be $4\text{–}6 \times 10^{30}$, comprising more than 106 different genospecies within more than 70 enormous phyla.

Whitman et al. (1998) reported microorganisms are the first living organisms on earth and play key role in global nutrient cycles such as carbon, nitrogen and sulphur cycles, and majority of the microorganisms reside in seawater, soil, and the sediment, each of these holding approximately $10^4\text{–}10^7$ cells/ml, $10^6\text{–}10^9$ cells/g and $10^5\text{–}10^8$ cells/cm$^3$, respectively. Apart from the genetic diversity described for these organisms, they are also a rich source for novel molecules such as antibiotics and biocatalysts.

## 3.2   SOIL AS A MICROBIAL HABITAT

Soil is probably the most challenging of all natural environments for microbiologists, with respect to the microbial community size and the diversity of species present. Robe et al. (2003) reported complexity of soil cause soil microbial diversity. Rajendhran and Gunasekaran (2008) found the complexity of microbial diversity results from multiple interacting parameters including pH, water content, soil structure, climate variations and biotic activity. Hassink et al. (1993) studied prokaryotes are the most abundant organisms in soil and can form the largest component of the soil biomass. Paul et al. (1989) and Richter et al. (1999) reported 1 g of forest soil contains an estimated $4 \times 10^7$ prokaryotic cells, whereas 1 g of cultivated soil and grasslands contain an estimated $2 \times 10^9$ prokaryotic cells. Despite the high concentration of organic matter in most soil types, only low concentration of organic carbon are readily available to microorganisms because the transformation of most of the organic matter that is derived from plants, animals and microorganisms into humus by the combination of microbiological and abiotic processes, and the uneven distribution of microorganisms and organic compounds in the soil matrix.

## 3.3   UNCULTURABLE MICROORGANISMS

Most of the soil microorganisms seem extremely well adapted to their environment; however, they cannot be cultured under the usual laboratory conditions. Amann et al. (1995) and Ferrer et al. (2005) reported majority of microbes on our planet have not been studied, largely because only less than 1% of them are culturable by conventional methods. Schloss and Handelsman1 studied that microorganisms present in many environments are not readily culturable and therefore not accessible for biotechnology. Handelsman et al. (2002) coined the term metagenomics, which refers to the genomic analysis of the collective microbial assemblage found in an environmental sample using an approach based either on heterologous expression or sequencing. DeLong and Karl (2005) and Sogin et al. (2006) reported the accessibility of unculturable microorganisms for their diversity, ecology and novel compounds can be overcome using metagenomics. Ferrerm et al. (2005) reported metagenomics has led to the accumulation of deoxyribonucleic acid (DNA) sequences, and these sequences are exploited for novel biotechnological applications. Due to presence of majority of unculturable microbes in soil, metagenome searches will always result in identification of hitherto unknown genes and proteins. Therefore, the probability of untapped hitherto unknown sequence makes metagenomics more favourable approach than searches in already cultured microorganisms. According to Sogin et al. (2006) natural diversity is known to be the best supplier of novel molecules like enzymes and antibiotics, which can be better explained by the vast majority of soil and other microbial niches. Torsvik et al. (2002) studied reassociation kinetics of DNA isolated from various soil samples; the number of distinct prokaryotic genomes has been estimated to range from 2000 to 18,000 genomes per gram of soil. Daniel (2005) intensively studied that microbial diversity depend on changes in the water content and other environmental factors such as pH, availability of oxygen or temperature. Soil environment is extremely diverse and soil microbes are exposed to extremes in pressure, temperature, salinity and nutrient availability. These distinct environmental niches are likely to possess highly diverse bacterial communities, possessing potentially unique novel compounds. Kennedy et al. (2008) reported that the microbial enzymes isolated from such environments are likely to have a range of diverse biochemical and physiological characteristics that have allowed the microbial communities to adapt and ultimately thrive in these conditions. Thus, the potential exists to exploit the enzymes produced by these soil microbes that are likely to possess unique biocatalytic activities capable of functioning under extreme conditions.

## 3.4   CULTURABLE MICROORGANISMS

Many culturable microorganisms such as bacteria, yeast and fungi are known to secret important metabolites. According to Handelsman, these culturable microorganisms require their natural environment in which they exist for survival and multiplication. Therefore, they produce industrially important biocatalysts, metabolites and antibiotics. Various physiological parameters are required to maintain their habitat; therefore more than 99% of microorganisms are unable to grow under laboratory conditions.

## 3.5   MICROBIAL SPECIES FOR BENEFICIAL PRODUCTS AND PROCESSES

Bacteria are life forms that are single celled and microscopic in size and exist in every habitat and participate in the Earth's element cycle. Microbes are involved in the production of oxygen, biocatalysts, bioactive compounds, biomass control and bioremediation. Some microbes also lead a symbiotic type of lifestyle in most multicellular organisms. Vuyst and Leroy (2007) reported that lactic acid bacteria show antibacterial activity due to production of organic acid and some other compounds such as bacteriocins and antifungal peptides. Wortman (1882), the first physiologist, described that living cells can adopt themselves to the utilization of certain food stuffs. He observed that bacterial species produce amylase when grown in the media containing starch. He also reported that yeasts produce invertase. Jacoby (1917) and Passmore et al. (1937) reported the production of urease by *Proteus vulgaris* when grown in media containing l-leucine or d-isoleucine. Lewis (1934) observed that *Escherichia coli* produce lactase.

## 3.6   METAGENOMICS

Schloss and Handelsman (2003) reported that more than 99% of the micro-organisms present in many natural environments are not readily cultur-able and, therefore, not accessible for biotechnology or basic research. Metagenomics is the culture-independent analysis of a mixture of microbial genomes using an approach based either on expression or on sequencing. The combined genomes of all the members present in a particular habitat build up the metagenome. Handelsman (2004) studied metagenomics that involves isolating DNA from an environmental sample, cloning the DNA into a suitable vector, transforming the clones into a host bacterium, and screening the resulting transformants for either phylogenetic markers or expression of specific traits, or they can be sequenced randomly (Entcheva et al., 2001).

Metagenomics is the genomic analysis of uncultivable microorganisms by direct extraction and cloning of DNA from environmental samples. Rondon et al. (2000) reported metagenomics term is derived from the statistical concept of meta-analysis (the process of statistically combining separate analysis and genomics that is the comprehensive analysis of an organism's genetic material). Ferrer et al. (2009) reported metagenomics has led to the accumulation of DNA sequences, and these sequences are exploited for novel biotechnological applications. Bott et al. (1969) used microscopic examination and staining technique with fluorescent antibodies raised against cultured microbes of the taxonomic groups suspected to inhabit the environment to study unculturable microbes in hot springs. Pace et al. (1986) studied ribosomal ribonucleic acid (rRNA) sequences for analysing natural microbial population. They used direct analysis of 5S and 16S rRNA gene sequences in the environment to describe the diversity of microorganisms in an environmental sample without culturing, thereby, highlighting the need for nontraditional techniques to understand the microbial world. According to He et al. (2007), metagenomics methodology has been developed as an effective tool for the discovery of new natural products and microbial functions.

## 3.7   SOIL METAGENOMICS FOR DESIRABLE GENES

Soil is considered as a complex environment, which appears to be a major reservoir of microbial diversity resulting from multiple interacting parameters, including pH, water content, soil structure, climate variations and biotic activity (Sleator et al., 2008). Most of the soil microorganisms seem extremely well adapted to their environment; however, they cannot be cultured under laboratory conditions. Only 0.1–1% of all microorganisms present in nature can be cultured under conventional laboratory conditions (Amann et al., 1995). The genomes of these mainly uncultured species encode a largely untapped reservoir of novel enzymes and metabolic capabilities. This leaves researchers unable to study more than 99% of microorganisms in some environments. Microorganisms have unique and potentially very useful abilities such as waste degradation or synthesis of compounds that could be useful in drugs or antibiotics. The approach has been used over a decade, however the term "metagenomics" was first used by Jo Handelsman in 2004, which was defined as "the application of modern genomics techniques to the study of communities of microbial organisms directly in their natural environments, bypassing the need for isolation and lab cultivation of individual species". Each organism in an environment has a unique set of genes in its genome; the combined genome of all the community members make up the "metagenome". Metagenomics is applied in many different research fields, particularly in microbial ecology, biodiversity and biotechnology (Handelsman et al., 2002). Metagenomics is the culture-independent analysis of a mixture of microbial genomes using an approach based either on expression or on sequencing (Riesenfeld et al., 2004). It is the study of metagenome, the genetic material recovered directly from environmental samples. The broad field may also be referred to as environmental genomics, ecogenomics or community genomics. The term "metagenomics" is derived from the statistical concept of meta-analysis and genomics to capture the notion of the analysis of a collection of similar but not identical items as in a meta-analysis which is an analysis of analyses (Entcheva et al., 2001).

Due to its ability to reveal the previously hidden diversity of microscopic life, metagenomics offers a powerful lens for viewing the microbial world that has the potential to revolutionize understanding of the entire living world. Due to overwhelming majority of non-culturable microbes in soil, metagenome searches will always result in identification of hitherto unknown genes and proteins. Thus, the probability of uncovering so far unknown sequence makes this approach more favourable than searches in already cultured microbes. Metagenomics allows the isolation of high quality DNA from variety of environments, that is contaminated subsurface sediments, ground water, surface water from rivers, marine picoplankton, soil, hot springs and mud holes in solfataric fields and so forth to study natural microbial communities without the need for cultivation. Metagenomics concerns the extraction, cloning, functional screening (Handelsman, 2004) and direct random shotgun sequencing of entire genetic complement of a habitat (Tyson et al., 2004); it is an approach that allows the investigation of the extensive diversity of individual genes and their products as well as analysis of entire operons encoding biosynthetic, metabolic or biodegradative pathways (DeLong et al., 2006). There is a vast amount of information held within the genome of uncultured micro-organisms and metagenomics is one of the key technologies used to access and investigate this potential. Considering the diversity of microbial species, the large population of soil microorganisms and the complex soil matrix, which contains many compounds, such as humic acids that bind to DNA and interfere with the enzymatic modification of DNA, recovery of microbial soil DNA that represents the resident microbial community and is suitable for cloning or polymerase chain reaction (PCR) is still an important challenge.

## 3.8   SOIL METAGENOMICS: TOOL FOR NOVEL COMPOUNDS

Soil is one of the most challenging and complex environment with respect to the microbial community size and the diversity of species present in it. Rajendhran and Gunasekaran (2008) reported that the complexity of microbial diversity in soil results from multiple parameters like soil structure, pH, water content, climatic variations and biotic activity. Soil represents the largest reservoir of microbial genomic and taxonomic diversity on the planet but the majority of these microbes are still uncharacterized and represent an enormous unexplored reservoir of genetic and metabolic diversity. Richter and Markewitz (1995) reported the estimated number of distinct prokaryotic genomes range from 2000 to 18,000 genomes per gram of soil using DNA–DNA reassociation kinetics from various soil samples. Tringe and Rubin (2005) reported the complexity of the bacterial DNA in any environment, as for instance soil, is at least 100 fold greater than others. Due to presence of major microbial diversity in soil, soil metagenomics is a good approach to study untapped microorganisms and their novel products.

Soil metagenomics includes the extraction of soil metagenome and subsequent construction and screening of clone libraries, which paved the way for the culture independent exploration of complex soil microbial communities for taxonomic study and recovery of novel biomolecules. The validity of the soil metagenomics is strongly dependent on efficient extraction of high quality microbial DNA and obtaining representative nucleic acids from entire microbial communities; which is a technological challenge. Singh et al. (*2013*) reported strategy for metagenomic DNA isolation and computational studies to remove humic acid. There are several research articles and reviews describing and assessing the standard proto-cols for efficient extraction of metagenome from different soil samples but none is universal for all soil types (Desai and Madamwar, 2010). Therefore, standardization of total DNA extraction technique is desirable according to the sample type as the composition of habitats varies with respect to their matrix, organic/inorganic compounds and biotic factors.

Metagenomics being a young and exciting technique has a broad scope of application in the field of biotechnology (Fig. 3.1). The current technology has proven to be sufficiently powerful to yield products for solving real world problems, including the discovery of new antibiotics and enzymes like amylase, protease, lipase, cellulase, amidase, oxidore-ductase, polyketide synthase, agarases and dehydratases. Recent studies in Sargasso Sea (Venter et al., 2004), acid mine drainage, soil and sunken whale skeletons have used the shotgun sequencing approach to sample the genomic content of these varied environments (Tringe et al., 2005). Metagenomics coupled with gene arrays, proteomics expression-based analysis and microscopy will provide insights into the studies of prob-lems such as genome evolution and members of particular niches that are currently hindered by our inability to culture most microorganisms in pure condition (Allen and Banfield, 2005).

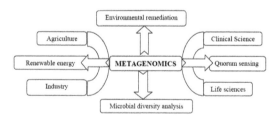

**FIGURE 3.1**    Practical applications of metagenomics.

Biotechnological applications from metagenomics may be fostered by the pursuit of fundamental ecological studies and focused screens for bioprospecting, just as both basic and applied approaches have contributed to the discovery of antibiotics and industrial enzymes from cultured micro-organisms. Novel genes and gene products discovered through metage-nomics include the first bacteriorhodopsin of bacterial origin, novel small molecules with antimicrobial activity and new members of known protein families, such as $Na^+(Li^+)/H^+$ antiporters, Rec A, DNA polymerase and antibiotic resistance determinants.

## 3.9   SEQUENCE-BASED METAGENOMICS

Sequence-based analysis (Fig. 3.2) can involve complete sequencing of clones containing phylogenetic anchors that indicate the taxonomic group which is the probable source of DNA fragment. Sequence-based screening

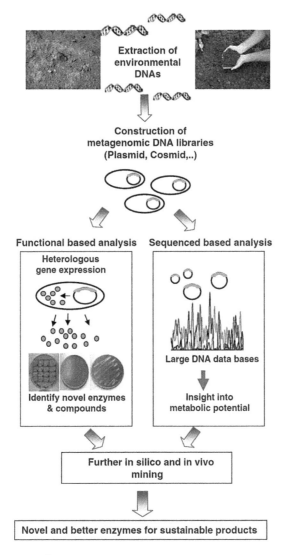

**FIGURE 3.2**   Approaches to metagenomics.

identifies the gene, genomic fragment or complete genome of interest through direct sequencing or sequence homology, for example, by hybridization or PCR amplification, and it is not dependent on the expression of cloned genes in heterologous host. Sequencing may either be random or targeted (Courtois et al., 2001). Target genes or pathways may contain genetic information which is ecologically or biotechnologically interesting. As a form of sequence-based screening, shotgun sequencing of metagenomic libraries has recently provided vast amount of data, including phylogenetic relationships, millions of novel genes, and deduced metabolic pathways of unculturable bacteria (Ferrer et al., 2009). The identification of a phylogenetic marker gene within a genomic fragment enables the linking of the sequence information, which could be a biologically interesting function, to particular phyla (Quinn et al., 1983). The alternative to a phylogenetic marker-driven approach is to sequence random clones, which has produced dramatic insights, especially when conducted on massive scale (Schmeisser et al., 2007). The distribution and redundancy of functions in a community, linkage of traits, genomic organization and horizontal gene transfer can all be inferred from sequence-based analysis (Quaiser et al., 2003). Different metagenomes have been sequenced from a large variety of environments such as soil, human gut, faeces and global oceans (Detter et al., 2002). Next generation sequencing techniques have been developed to allow large-scale analysis of microbial niches resulting in novel applications such as comparative community metagenomics, metatranscriptomics and metabolomics (Chistoserdova, 2010). The reconstruction of genomes of uncultured organisms in a community in acid mine drainage illustrates the power of large-scale sequencing of uncultured communities.

The sequence of flanking regions of a metagenomic DNA has revealed a bacteriorhodopsin like gene. Its gene product was shown to be an authentic photoreceptor, leading to the insight that bacteriorhodopsin genes are not limited to Archaea but is in fact abundant among the *proteobacteria* of the ocean (Beja et al., 2000). These studies have made new linkages between phylogeny and function, indicating the surprising abundance of certain types of genes and reconstruction of the genomes of unculturable organisms.

## 3.10  FUNCTION-DRIVEN METAGENOMICS

Function-based analysis (Fig. 3.2) enables identification of new enzymes, antibiotics or other metabolites in libraries from diverse environments.

Function-driven approaches include three different ways to recover novel biomolecules:

1.  Phenotypical detection of the desired activity (Beloqui et al., 2010)
2.  Heterologous complementation of host strains or mutants (Chen et al., 2010)
3.  Induced gene expression (Uchiyama and Miyazaki, 2010)

Phenotypical detection utilizes chemical dyes and insoluble or chro-mophore-bearing derivatives of enzyme substrates incorporated into the growth medium, where they register the specific metabolic capabilities of individual clones.

A different category of function-driven approach is based on heterol-ogous complementation of host strains or mutants of host strains which requires the targeted genes for growth under selective conditions. This tech-nique allows a simple and fast screening of complex metagenomic libraries comprising millions of clones. The scope of this approach is to access complete genes and metabolic pathways without any prior knowledge of sequence information of the target gene. This enables discoveries of novel and previously unknown genes and gene products. One major limitation to heterologous expression is that the host must have a compatible expres-sion system for the cloned environmental DNA. Therefore, the frequency of detected activities is often low, which in turn requires the use of high throughput screening systems. However, the development of new vectors and expression hosts and the fact that DNA from bacteria of some phylo-genetic groups is compatible with the most commonly used expression system, such as *E. coli,* further broadens the scope of the technology (Wang et al., 2000). Induced gene expression is also termed as substrate-induced gene expression screening (SIGEX). This high-throughput screening approach employs an operon trap *gfp*-expression vector in combination with fluorescence-activated cell sorting. The screen is based on the fact that catabolic-gene expression is induced mainly by specific substrates and is often controlled by regulatory elements located close to catabolic genes. To perform SIGEX, metagenomic DNA is cloned upstream of the *gfp* gene, thereby placing the expression of *gfp* under the control of promoters present in the metagenomic DNA. Clones influencing *gfp* expression upon addition of the substrate of interest are isolated by fluorescence-activated cell sorting. One drawback of this approach is the possible activation of transcriptional regulators by effectors other than the specific substrates (Galvao et al., 2005). A similar type of screening designated metabolite-regulated expression (METREX) is to identify metagenomic clones producing small molecules; a biosensor that detects small diffusible signal molecules that induce quorum sensing is inside the same cell as the vector harbouring a metagenomic DNA fragment. Another screen based on induced gene expression, termed product induced gene expression (PIGEX). In this reporter assay system, enzymatic activities are also detected by the expression of *gfp*, which is triggered by the product formation.

## 3.11   METAGENOMIC LIBRARY CONSTRUCTION AND ITS ANALYSIS

Two parameters are important for metagenomic library construction, first is the large size of the metagenome, and such diversity requires improved cloning efficiency so that the clones in the gene library can represent the entire metagenome. Second, the size and cluster organization of the genes are involved in the synthesis of secondary metabolites. In principle, screens of soil-based libraries can be based either on metabolic activity (function-driven approach) or on nucleotide sequence.

Polymerase chain reaction is most commonly used for sequence-driven screening of soil-based libraries or soil DNA. Hybridization using target-specific probes has also been used to screen soil-based libraries. This approach quickly identifies clones that have potential applications in medicine, agriculture or industry by focusing on natural products or proteins that have useful activities. Both approaches require suitable primers and probes that are derived from conserved regions of known genes and gene products, so applicability is limited to the identification of new members of known gene families. This approach has been used to identify phylogenetic anchors such as 16S rRNA genes encoding enzymes with highly conserved domains such as polyketide synthases, gluconic acid reductases and nitrile hydratases. Random sequencing of soil-derived libraries is another approach to characterize the soil ecosystem on a genomic level, but the species richness of soil habitats would require enormous sequencing and assembly efforts.

Most of the screening methods to isolate gene or gene clusters for novel biocatalysts or small molecule are based on detecting activity from library containing clones. This is the only strategy that has the potential to identify new classes of genes that encode either known or new functions. This approach has been used to clone genes from soil communities that code for lipases, proteases, oxidoreductases, amylases, cellulases, antibiotics, antibiotic resistance enzymes and membrane proteins. Function-driven approaches can include screening of metagenomic libraries for a specific function. Media containing substrates such as tetrazolium chloride, Schiff reagent, tributyrin and skimmed milk have been used for identifying clones exhibiting genes for 4-hydroxybutyrate utilization, carbonyls production, and lipolytic, proteolytic activity, respectively. Chemical dyes and insoluble or chromophore-bearing derivatives of enzyme substrates can be incorporated into the growth medium solidified with agar to monitor enzymatic functions of individual clones.

## 3.12   APPROACHES TO METAGENOMICS

Metagenomic analysis involves isolation of DNA from environmental samples, cloning the DNA/gene into suitable vector, transferring the clones into the host bacterium and screening of the resultant transformants. The clones can be screened for phylogenetic markers or "anchors" such as 16S rRNA and rec A, or for other conserved genes by hybridization or multiplex PCR or for expression of specific traits, such as enzyme activity or antibiotic production, or they can be sequenced randomly (Diaz-Torres et al., 2003).

Each approach has strengths and limitations; together these approaches have enriched our understanding of the unculturable microorganisms. The metagenomics approach includes both sequence and function-based analysis of DNA extracted directly from environment. Metagenomics comprises two basic approaches: sequence-based analysis and functional approach.

### 3.12.1   SEQUENCE- AND FUNCTION-BASED ANALYSIS

Daniel and Simon (2009) reported sequence-based screening involves direct sequencing of metagenomic DNA, either with or without cloning prior to sequencing and then subjecting the sequences to bioinformatics analysis. However, sequence-based analysis may also be said to involve complete sequencing of clones containing phylogenetic anchors like 16S rRNA, indicating the taxonomic group that is the probable source of the DNA fragments. Alternatively, random sequencing can be conducted, and once a gene of interest is identified, phylogenetic anchors can be sought in the flanking DNA to provide a link of phylogeny with the functional gene. A new approach in this field is PCR-denaturing gradient gel electrophoresis (DGGE) followed by metagenomic walking.

Beja et al. (2000) reported the sequence of flanking DNA revealed a bacteriorhodopsin-like gene. Its gene product was shown to be an authentic photoreceptor, leading to the insight that bacteriorhodopsin genes are not limited to Archaea but is in fact abundant among the *proteobacteria* of the ocean. Novel enzymes such as chitinases, alcohol oxidoreductases, diol dehydratases and enzyme conferring antibiotic resistance can be recovered by sequence-driven approaches. Venter et al. (2004) applied random shotgun sequencing approach first time for the Sargasso Sea, which involves the direct cloning of metagenome without prior sequence knowledge.

Banik and Brady (2008) isolated two novel glycopeptides encoding gene clusters from desert soil by a PCR-based screening, which can serve as substitutes of currently used antibiotics such as vancomycin. Metagenomics has identified genes encoding polyketide synthases (PKSs) and peptide synthetases, which contribute to synthesis of complex antibiotics. Various authors have reported biocatalysts using functional-based analysis from metagenome.

## REFERENCES

Allen, E. E.; Banfield, J. F. Community Genomics in Microbial Ecology and Evolution. *Nat. Rev. Microbiol.* **2005,** *3,* 489–498.

Amann, R.; Ludwig, W.; Schleifer, K. Phylogenetic Identification and in situ Detection of Individual Microbial Cells Without Cultivation. *Microbiol. Rev.* **1995,** *59,* 143–169.

Banik, J. J.; Brady, S. F. Cloning and Characterization of New Glycopeptide Gene Clusters Found in an Environmental DNA Megalibrary. *Proc. Natl. Acad. Sci. U. S. A.* **2008,** *105,* 17273–17277.

Beja, O., et al. Bacterial Rhodopsin: Evidence for a New Type of Photography in the Sea. *Science.* **2000,** *289,* 1902–1906.

Beloqui, A., et al. Diversity of Glycosyl Hydrolases from Cellulose-Depleting Communities Enriched from Casts of Two Earthworm Species. *Appl. Environ. Microbiol.* **2010,** *76,* 5934–5946.

Bott, T. L.; Brock, T. D. Bacterial Growth Rates Above 90°C in Yellow Stone Hot Springs. *Science.* **1969,** *164,* 1411–1412.

Chen, I. C., et al. Lysine Racemase: A Novel Non-antibiotic Selectable Marker for Plant Transformation. *Plant Mol. Biol.* **2010,** *72,* 153–169.

Chistoserdova, L. Recent Progress and New Challenges in Metagenomics for Biotechnology. *Biotechnol. Lett.* **2010,** *32,* 1351–1359.

Courtois, S., et al. Quantification of Bacterial Subgroups in Soil: Comparison of DNA Extracted Directly from Soil or from Cells Previously Released by Density Gradient Centrifugation. *Environ. Microbiol.* **2001,** *3,* 431–439.

Courtois, S., et al. Recombinant Environmental Libraries Provide Access to Microbial Diversity for Drug Discovery from Natural Products. *Appl. Environ. Microbiol.* **2003,** *69,* 49–55.

Daniel, R. The Metagenomics of Soil. *Nat. Microbiol.* **2005,** *3,* 470–478.

Daniel, R.; Simon, C. Achievements and New Knowledge Unravelled by Metagenomic Approaches. *Appl. Microbiol. Biotechnol. Lett.* **2009,** *85,* 265–276.

DeLong, E. F.; Karl, D. M. Genomic Perspectives in Microbial Oceanography. *Nature,* **2005,** *437,* 336–342.

DeLong, E. F., et al. Community Genomics Among Stratified Microbial Assemblages in the Ocean's Interior. *Science.* **2006,** *311,* 496–503.

Desai, C.; Madamwar, D. Extraction of Inhibitor-free Metagenomic DNA from Polluted Sediments, Compatible with Molecular Diversity Analysis Using Adsorption and Ion-exchange Treatments. *Bioresour. Technol.* **2010,** *98,* 761–768.

Detter, J. C. et al. Isothermal Strand-displacement Amplification Applications for High-Throughput Genomics. *Genomics.* **2002,** *80,* 691–698.

Diaz-Torres, M. L., et al. Novel Tetracycline Resistance Determinant from the Oral Metagenome. *Antimicrob. Agents Chemother.* **2003,** *47,* 1430–1432.

Entcheva, P., et al. Direct Cloning from Enrichment Cultures, a Reliable Strategy for Isolation of Complete Operons and Genes from Microbial Consortia. *Appl. Environ. Microbiol.* **2001,** *67,* 89–99.

Ferrer, M., et al. Metagenomics for Mining New Genetic Resources of Microbial Communities. *J. Mol. Microbiol. Biotechnol.* **2009,** *16,* 109–123.

Ferrerm M.; Martinez-Abarcam, F.; Golyshin, P. N. Mining Genomes and Metagenomes for Novel Catalysts. *Curr. Opin. Biotechnol.* **2005,** *16,* 588–593.

Galvao, T. C.; Mohn, W. W.; de Lorenz, V. Exploring the Microbial Biodegradation and Biotransformation Gene Pool. *Trends Biotechnol.* **2005,** *23,* 497–506.

Handelsman, J. Metagenomics: Application of Genomics to Uncultured Microorganisms. *Microbiol. Mol. Biol. Rev.* **2004,** *68,* 669–685.

Handelsman, J., et al. Cloning the Metagenome: Culture-independent Access to the Diversity and Functions of the Uncultivated Microbial World. In *Methods in Microbiology: Functional Microbial Genomics;* Wren B., Dorrell N., Eds.; Academy press: New York, 2002.

Hassink, J., et al. Relationship Between Habitable Pore Space, Soil Biota, and Mineralization Rates in Grassland Soil. *Soil. Biol. Biochem.* **1993,** *25,* 47–55.

He, J. Z., et al. Methodology and Application of Soil Metagenomics. *Chin. Acad. Sci.* **2007,** *18,* 212–218.

Jacoby, M. Uber Fermentbildung. *Biochem. Z.* **1917,** *81,* 332–341.

Kennedy, J.; Marchesi, J. R.; Dobson, A. D. W. Marine Metagenomics: Strategies for the Discovery of Novel Enzymes with Biotechnological Applications from Marine Environments. *Microb. Cell Fact.* **2008,** *7,* 27.

Lewis, I. M. Bacterial Variation with Special Reference to Behavior of Some Mutabile Strains of Colon Bacteria in Synthetic Media. *J. Bacteriol.* **1934,** *28,* 619–639.

Pace, N. R., et al. The Analysis of Natural Microbial Populations by Ribosomal RNA Sequences. *ASM News.* **1985,** *51,* 4–12.

Pace, N. R., et al. Analysing Natural Microbial Populations by rRNA Sequences. *Adv. Microb. Ecol.* **1986,** *9,* 1–55.

Passmore R.; Yudkin, J. The Effect of Carbohydrates and Allied Substances on Urease Production by *Proteus vulgaris. Biochem. J.* **1937,** *31,* 318–322.

Paul, E. A.; Clark, F. E. *Soil Microbiology and Biochemistry.* Academic Press: San Diego, 1989.

Quaiser, A., et al. Acidobacteria from a Coherent but Highly Diverse Group within the Bacterial Domain: Evidence from Environmental Genomics. *Mol. Microbiol.* **2003,** *50,* 563–575.

Quinn, D.; Shirai, K.; Jackson, R. L. Lipoprotein Lipase: Mechanism of Action and Role in Lipoprotein Metabolism. *Prog. Lipid Res.* **1983,** *22,* 35–78.

Rajendhran, J.; Gunasekaran, P. Strategies for Accessing Soil Metagenome for Desired Applications. *Biotechnol. adv.* **2008,** *26,* 576–590.

Richter. D. D.; Markewitz, D. How Deep is Soil? *BioScience.* **1995,** *45,* 600–609.

Richter, L., et al., Methanogenic Archaea and CO2-dependent Methanogenesis on Washed Rice Roots. *Environ. Microbiol.* **1999,** *1,* 159–166.

Riesenfeld, C. S.; Goodman, R. M.; Handelsman, J. Uncultured Soil Bacteria are a Reservoir of New Antibiotic Resistance Genes. *Environ. Microbiol.* **2004,** *6,* 981–989.

Robe, P., et al. Extraction of DNA from Soil. *Eur. J. Soil Biol.* **2003,** *39,* 183–190.

Rondon, M. R., et al. Cloning and Soil Metagenome: a Strategy for Accessing Thegenetic and Functional Diversity of Uncultured Microorganisms. *Appl. Environ. Microbiol.* **2000,** *66,* 2541–2547.

Schloss, P. D.; Handelsman, J. Biotechnological Prospects from Metagenomics. *Curr. Opin. Biotechnol.* **2003,** *14,* 303–310.

Schmeisser, C.; Steele, H.; Streit, W. R. Metagenomics, Biotechnology with Non-culturable Microbes. *Appl. Microbiol. Biotechnol.* **2007,** *75,* 955–962..

Singh, S. P.; Sagar, K.; Konwar, B. K. Strategy in Metagenomic DNA Isolation and Computational Studies of Humic Acid. *Curr. Res. Microbiol. Biotechnol.* **2013,** *1,* 9–11.

Sleator, R. D.; Shortall, C.; Hill, C. Metagenomics. *Lett. Appl. Microbiol.* **2008,** *47,* 361–366.

Sogin, M. L., et al. Microbial Diversity in the Deep Sea and the Underexplored—Rare Biosphere. *Proc. Natl. Acad. Sci. U. S. A.* **2006,** *103,* 12115–12120.

Torsvik, V., et al. Microbial Diversity and Function in Soil: from Genes to Ecosystems. *Curr. opin. Microbiol.* **2002,** *5,* 240–245.

Tringe S. G.; Rubin, E. M. Metagenomics: DNA Sequencing of Environmental Samples. *Nat. Rev. Genet.* **2005,** *6,* 805–814.

Tringe, S. G., et al. Comparative Metagenomics of Microbial Communities. *Science.* **2005,** *308,* 554–557.

Tyson, G. W., et al. Community Structure and Metabolism Through Reconstruction of Microbial Genomes from the Environment. *Nature*. **2004,** *428,* 37–43.

Uchiyama, T.; Miyazaki, K. Product-Induced Gene Expression (PIGEX): A Productresponsive Reporter Assay for Enzyme Screening of Metagenomic Libraries. *Appl. Environ. Microbiol.* **2010,** *76,* 7029–7035.

Venter, J. C., et al. Environmental Genome Shotgun Sequencing of the Sargasso Sea. *Science.* **2004,** *304,* 66–74.

Vuyst, L. D.; Leroy, F. Bacteriocins from Lactic Acid Bacteria: Production, Purification and Food Applications *J. Mol. Microbiol. Biotechnol.* **2007,** *13,* 194–199.

Wang, G. Y. S., et al. Novel Natural Products from Soil DNA Libraries in a Streptomycete Host. *Org. Lett.* **2000,** *2,* 2401–2404.

Whitman, W. B.; Coleman, D. C.; Wiebe, W. J. Prokaryotes: the Unseen Majority. *Proc. Natl. Acad. Sci. U. S. A.* **1998,** *95,* 6578–6583.

Wortman, J. Untersuchungen Ilber Das Diastatische Ferment der Bakterien. *Z. physiol. Chem.* **1882,** *6,* 287–329

# CHAPTER 4

# ACCESSING METAGENOMICS

## CONTENTS

### 4.1   METHODS

There are a range of methods that are available for extraction of metagenomic deoxyribonucleic acid (DNA) from environmental samples. However, none of the methods reported hitherto is universally applicable and every type of soil sample requires optimization of DNA extraction methods (Zhou et al., 1996). Metagenomic DNA isolation from soil and sediment samples can be broadly classified into direct and indirect extraction procedures. Direct DNA isolation is based on cell lysis within the sample matrix and subsequent separation of DNA from the matrix and cell debris or separation of the cells from the soil matrix followed by cell lysis (Ogram et al., 1987). Since no single method of cell lysis is appropriate for all types of soils, different combinations and modifications of lysis protocols may be needed for different soil samples (Hurt et al., 2001).

   The indirect approach involves the separation of cells from the soil matrix followed by cell lysis and DNA extraction (Holben et al., 1988), Gabor et al. (2003) and Courtois et al. (2001) reported that DNA extraction methods based on cell separation, although less efficient in terms of the amount of DNA recovered, are less harsh than direct lysis methods. DNA recovered by direct method seems to be less contaminated with matrix compounds, including humic substances. In addition, the average size of the isolated DNA is larger than that typically obtained by the direct lysis approach and is therefore more suitable for the generation of large-insert libraries. After cell lysis, organic solvents (phenol, phenol–chloroform, chloroform–isoamyl alcohol) and saturated salt solutions (sodium chloride, ammonium acetate, potassium acetate and sodium acetate) have been used for deproteinisation.

The pH of the extraction buffer also plays a vital role in the recovery of soil metagenomic DNA. Frostegard et al. (1999) reported pH 9.0 as the optimum pH of the extraction buffer. DNA precipitation, which is performed to discard the extraction buffer and contaminants, is also a crucial step influencing the quality of metagenomic DNA. Few reports suggested that alcoholic precipitation favours the co-precipitation of humic acids, while polyethylene glycol (PEG) greatly reduces the humic substance co-precipitation. Five percent PEG yields significantly less humic acids without affecting polymerase chain reaction (PCR) so 5% PEG has been recommended for the precipitation of soil metagenomic DNA (Porteous et al., 1997).

Humic acid is a principal component of humic substances, which are the major organic constituents of soil, peat, coal, many upland streams, dystrophic lakes and ocean water (Arbeli and Fuentes, 2007). Humic acid is a complex mixture of many different acids containing carboxyl and phenolate groups so that the mixture behaves functionally as a dibasic acid or, occasionally, as a tribasic acid. A typical humic substance is a mixture of many molecules, some of which are based on a motif of aromatic nuclei with phenolic and carboxylic substituents, linked together. Humic substances are structurally complex, polyelectrolytic, yellow to dark brown in colour and have the molecular mass range of 0.1 to more than 300 kDa. The humic substances have ability to bind and absorb water, ions and organic molecules. In addition, due to physicochemical properties similar to that of nucleic acids, humic substances along with the adsorbed organic molecules are generally co-extracted with DNA and affect almost all molecular biological methods such as hybridization, restriction digestions of DNA, PCR and bacterial transformation. Humic compounds absorb at 230 nm, however, estimation of humic acids is influenced by the concentration of nucleic acids and protein contaminants with the absorbance in ultraviolet (UV) range. Therefore, absorbance at

320 nm can be used to measure the level of humic acids, which is independent of DNA and protein content. Miller et al. (2001) and Howeler et al. (2003) have reported two methods of measuring humic acid levels: (i) absorbance at 340 nm and (ii) fluorescence excitation at 471 nm and emission at 529 nm. Spectrophotometric quantification of soil DNA is challenging due to the presence of humic substances, which interferes with the traditional estimation ($A_{260}$). Estimation of DNA concentration by the densitometric analysis of ethidium bromide stained Agarose gel is another major approach. Alternatively, fluorometric analysis using the fluorescent dye PicoGreen offers efficient quantification of soil DNA (Jackson et al., 1997). Successful PCR amplification is generally used as an indicator of soil DNA purity. The clone ability of metagenomic DNA can be studied by restriction digestion and ligation efficiencies. Purity of metagenomic DNA can be estimated in terms of blunt end cloning efficiencies in a blunt end restriction site such as *Eco*RV of a high copy vector pUC-18 (Frost-egard et al., 1999).

Caesium chloride density gradient centrifugation is a widely used and efficient strategy for the purification of DNA from contaminants. However, due to the longer processing time, this method is not suitable for purification of multiple samples (Smalla et al., 1993). The addition of hexadecyl trimethyl ammonium bromide (CTAB) or polyvinylpolypyrrolidone to soil–buffer slurry before cell lysis inhibited the co-precipitation of humic substances. Agarose gel purification, electroelution and various chromatographical separations techniques viz. gel filtration, ion exchange, adsorption and so forth have also been used to purify DNA. Pretreatment of samples with calcium carbonate ($CaCO_3$) or purification of extracted DNA with calcium chloride ($CaCl_2$) has, however, been reported to be more efficient than commercial kits thereby, yielding PCR-compatible soil metagenomic DNA (Sagova-Mareckova et al., 2008).

## 4.2   EXPRESSION VECTOR/HOST SYSTEM

The expression vector varies with the range of target insert DNA and size of gene required for cloning. For small target genes, plasmids or lambda expression vectors such as pET-28a, pET-30b pET-32a and so forth are used for constructing DNA fragment libraries which are further screened for enzyme expression into suitable media. Large target genes having size range between 20 and 40 kb require expression libraries in cosmids and fosmids and up to 100–200 kb in bacterial artificial chromosome vectors.

Many genes from environmental samples may not be expressed efficiently in heterologous host due to difference in G+C contents, codon usage, transcription and/or translation initiation signals, protein-folding elements, post translational modifications, such as glycosylation, or toxicity of the active enzyme. According to Vakhlu et al. (2012) selection of suitable vector system and expression host containing appropriate transcription and translation-initiation sequences can reduce this limitation. Although common *Escherichia coli* host strains are not compatible with many environmental genes. Novagen, USA developed an expression host *E. coli* Rosetta, which contain the tRNA genes for rare amino acid codons or co-expression of the chaperone proteins, such as GroES, GroEL, and heat-shock proteins. Li et al. (2009) reported the suitability of bacterial hosts such as *Pseudomona putida*, *Streptomyces lividans* and *Bacillus subtilis* for heterologous gene expression in *E. coli* Rosetta. Also apart from *E. coli*, the most suitable host, other bacteria like *Streptomyces* and *Pseudomonas* strains have also been used to expand the range of soil-derived genes.

## 4.3   FUNCTIONAL SCREENING OF METAGENOMIC LIBRARIES

Functional screening of metagenomic libraries can be done by chemical dyes and insoluble or chromophore-containing derivative medium (Table 4.1). Lee et al. (2007) and Waschkowitz et al. (2009) reported detection of the recombinant *E. coli* clones exhibiting protease activity on indicator agar containing skimmed milk as protease substrate. Hardeman and Sjoling (2007) and Lee et al. (2006) reported detection of lipolytic activity of recombinants by employing indicator agar containing tributyrin or tricaprylin as enzyme substrates. Recombinants with proteolytic and lipolytic activity are identified by formation of a clearing zone on solidified indicator medium.

**TABLE 4.1**   Functional Screening of Metagenomic Libraries for Antibiotics, Industrially Important.

| Gene | Habitat | Library type | Av. Insert (kb) | No. of clones screened | Substrate | Positive clones |
|---|---|---|---|---|---|---|
| Esterase/lipase (Khan et al., 2013) | Soil | Fosmid | 40 | 45,000 | *p*-nitro-phenyl ester | 4 |
| Esterase/lipase (Jiang et al., 2012) | Sediment | Fosmid | 36 | 20,000 | *p*-nitro-phenyl ester | 12 |
| Esterase/lipase (Jin et al., 2012) | Soil | Fosmid | 35 | 33,700 | *p*-nitro-phenyl ester | 8 |
| Esterase/lipase (Henne et al., 2000) | Soil | Plasmid | 6 | 2,86,000 | Tributyrin | 3 |
| Esterase/lipase (Entcheva et al., 2001) | Soil | BAC | 27 | 3648 | Bacto Lipid | 2 |
| Esterase/lipase (Castro et al., 2011) | Soil | Plasmid | 8 | 150,000 | Tributyrin | 10 |
| Esterase/lipase (Hong et al., 2007) | Forest soil | Fosmid | 40 | 31,000 | Tributyrin | 7 |
| Esterase/lipase (Hong et al., 2007) | Forest soil | Fosmid | 40 | 31,000 | Tributyrin | 7 |
| Esterase/lipase (Castro et al., 2011) | Soil | Fosmid | 35 | 65,000 | Tributyrin | 12 |
| Esterase/lipase (Brady et al., 2004) | Forest soil | Fosmid | 35 | 33,700 | Tributyrin | 8 |
| Lipase (Cieslinski et al., 2009) | Antarctic soil | Plasmid | 5 | 1000 | Olive oil | 1 |
| Lipase (Jimenez et al., 2012) | Forest soil | Plasmid | 4.6 | 20,000 | Tributyrin | 2 |
| Lipase (Jimenez et al., 2012) | Sandy soil | Fosmid | 40 | 25,344 | Tributyrin | 1 |

## REFERENCES

Arbeli, Z.; Fuentes, C. L. Improved Purification and PCR Amplification of DNA from Environmental Samples. *FEMS Microbiol. Lett.* **2007,** *272,* 269–275.

Brady, S. F.; Chao, C. J.; Clardy, J. Long-chain N-acyltyrosine Synthases from Environmental DNA. *Appl. Environ. Microbiol.* **2004,** *70,* 46865–46870.

Castro, A. P. de, et al. Construction and Validation of Two Metagenomic DNA Libraries from Cerrado Soil with High Clay Content. *Biotechnol. Lett.* **2011,** *33,* 2169–2175.

Cieslinski, H., et al. Identification and Molecular Modeling of a Novel Lipase from an Antarctic Soil Metagenomic Library. *Pol. J. Microbiol.* **2009,** *58*(3), 199–204.

Courtois, S., et al. Quantification of Bacterial Subgroups in Soil: Comparison of DNA Extracted Directly from Soil or from Cells Previously Released by Density Gradient Centrifugation. *Environ. Microbiol.* **2001,** *3,* 431–439.

Entcheva, P., et al. Direct Cloning from Enrichment Cultures, a Reliable Strategy for Isolation of Complete Operons and Genes from Microbial Consortia. *Appl. Environ. Microbiol.* **2001,** *67,* 89–99.

Frostegard, A., et al. Quantification of Bias Related to the Extraction of DNA Directly from Soils. *Appl Environ. Microbiol.* **1999,** *65,* 5409–5420.

Gabor, E. M.; de Vries, E. J.; Janssen, D. B. Efficient Recovery of Environmental DNA for Expression Cloning by Indirect Methods. *FEMS Microbiol. Ecol.* **2003,** *44,* 153–163.

Hardeman, F.; Sjoling, S. Metagenomic Approach for the Isolation of a Novel Low-Temperature- Active Lipase from Uncultured Bacteria of Marine Sediment. *FEMS Microbiol. Ecol.* **2007,** *59,* 524–534.

Henne, A., et al. Screening of Environmental DNA Libraries for the Presence of Genes Conferring Lipolytic Activity on *Escherichia coli*. *Appl. Environ. Microbiol.* **2000,** *66,* 3113–3116.

Holben, W. E., et al. DNA Probe Method for the Detection of Specific Microorganisms in the Soil Bacterial Community. *Appl. Environ. Microbiol.* **1988,** *54,* 703–711.

Hong, K. S., et al. Selection and Characterization of Forest Soil Metagenome Genes Encoding Lipolytic Enzymes. *J. Microbiol. Biotechnol.* **2007,** *17*(10), 1655–1660.

Howeler, M., et al. A Quantitative Analysis of DNA Extraction and Purification from Compost. *J. Microbiol. Methods.* **2003,** *54,* 37–45.

Hurt, R. A., et al. Simultaneous Recovery of RNA and DNA from Soils and Sediments. *Appl. Environ. Microbiol.* **2001,** *67,* 4495–4503.

Jackson, C. R., et al. A Simple, Efficient Method for the Separation of Humic Substances and DNA from Environmental Samples. *Appl. Environ. Microbiol.* **1997,** *63,* 4993–4995.

Jiang, X., et al. Identification and Characterization of Novel Esterases from a Deep-Sea Sediment Metagenome. *Arch. Miocrobiol.* **2012,** *194,* 207–214.

Jimenez, D. J., et al. A Novel Cold Active Esterase Derived from Colombian High Andean Forest Soil Metagenome. *World J. Microbiol. Biotechnol.* **2012,** *28,* 361–370.

Jin, P., et al. Overexpression and Characterization of a New Organic Solvent-Tolerent Esterase Derived from Soil Metagenomic DNA. *Bioresour. Technol.* **2012,** *116,* 234–240.

Lee D. G., et al. Screening and Characterization of a Novel Fibrinolytic Metalloprotease from a Metagenomic Library. *Biotechnol. Lett.* **2007,** *29,* 465–472.

Lee M. H., et al. Isolation and Characterization of a Novel Lipase from a Metagenomic Library of Tidal Flat Sediments: Evidence for a New Family of Bacterial Lipases. *Appl. Environ. Microbiol.* **2006,** *72,* 7406–7409.

Li Y. Z., et al., Cloning and Heterologous Expression of a new 3'-hydroxylase Gene from Lycoris Radiate. *Z Naturforsch C.* **2009,** *64,* 138–42.

Miller, D. N. Evaluation of Gel Filtration Resins for the Removal of PCR-Inhibitory Substances from Soils and Sediments. *J. Microbiol. Methods.* **2001,** *44,* 49–58.

Ogram, A.; Sayler, G. S.; Barkay, T. The Extraction and Purification of Microbial DNA from Sediments. *J. microbiol. methods.* **1987,** *7,* 57–66,.

Porteous L. A., et al. An Improved Method for Purifying DNA from Soil for Polymerase Chain Reaction Amplification and Molecular Ecology Applications. *Mol. Ecol.* **1997,** *6,* 787–791.

Sagova-Mareckova, M., et al. Innovative Methods for Soil DNA Purification Tested in Soils with Widely Differing Characteristics. *Appl. Environ. Microbiol.* **2008,** *74,* 2902–2907.

Smalla, K., et al. Rapid DNA Extraction Protocol from Soil for Polymerase Chain Reaction Mediated amplification. *J. Appl. Bacteriol.* **1993,** *74,* 78–85.

Vakhlu, J.; Ambardar, S.; Johri, B. N. Microorganisms in Sustainable Agriculture and Biotechnology. Springer: New York, 2012, 263–294.

Waschkowitz T; Rockstroh S.; Daniel, R. Isolation and Characterization of Metalloproteases with a Novel Domain Structure by Construction and Screening of Metagenomic Libraries. *Appl Environ. Microbiol.* **2009,** *75,* 2506–2516.

Zhou, J.; Bruns, M. A.; Tiedje, J. M. DNA Recovery from Soils of Diverse Composition. *Appl. Environ. Microbiol.* **1996,** *62,* 316–322.

# CHAPTER 5

# METAGENOMICS FOR LIPASE

## CONTENTS

Metagenomics is an approach to discover novel lipases. Several genes encoding metagenomic lipases have been identified in metagenomic libraries prepared from various environmental samples. Henne et al. (2000) reported a gene conferring lipolytic activity after screening of soil deoxyribonucleic acid (DNA) libraries. Brady et al. (2004) reported a long-chain N-acyl-tyrosine synthases from environmental DNA. Cieslinski et al. (2009) reported identification and molecular modelling of a novel lipase from an Antarctic soil metagenomic library. Jimenez et al. (2012) reported a novel cold active esterase derived from Colombian high Andean forest soil metagenome. Hong et al. (2007) characterized the forest soil metagenome for lipase encoding genes. Castro et al. (2011) constructed metagenomic DNA (mgDNA) libraries from Cerrado soil for untapped lipase genes. Lee et al. (2006) isolated and sequenced a novel lipase-encoding gene, *lipG*, from a tidal flat-derived metagenomic library. Zhang et al. (2009) identified a novel esterase gene EstCE1 from China soil. Esterase EstCE1 displays remarkable characteristics that cannot be related to the original environment from which they were derived. Kennedy et al. (2008) reported the strategies to discover novel biocatalysts with biotechnological applications from various terrestrial environmental niches. Zhang *et al.* (2009) cloned and characterized a novel esterase EstAS from activated sludge metagenome. EstAS consists of 834 bp and encodes a polypeptide of 277 amino acid residues with a molecular mass of 31 kDa.

This EstAS had optimal temperature and pH at 35°C and 9.0, respectively. Jiang et al. (2012) identified and characterized a novel esterase Est6 from a deep-sea sediment metagenome. Est6 consisted of 909 bp that encoded 302 amino acid residues. Est6 was most similar to a lipolytic enzyme from uncul-turable bacterium (ACL67845, 61% identity) isolated from the South China Sea marine sediment metagenome. Est6 revealed a cold-active esterase and exhibited highest activity at 20°C and 7.5 pH. Jin et al. (2012) reported over-expression and characterization of a new organic solvent-tolerant esterase EstC23, derived from soil mgDNA. EstC23 showed optimal activity at 40°C and retained about 50% maximal activity at 5–10°C.

## 5.1   COLLECTION OF ENVIRONMENTAL SAMPLES

Chemicals and reagents including the required microbial culture media have to be of analytical grade and purchased from the reputed companies and suppliers as shown in Table 5.1.

**TABLE 5.1**  Microbial Strains with Plasmids, Chemicals and Equipment, Bacterial Species/Strains, Plasmids with Their Genotypic Description and Source.

| Strains/plasmids | Genotype/description | Source |
|---|---|---|
| *Escherichia coli* DH10B | F-endA1 recA1 galE15 galK16 nupGrpsL ΔlacX74 | Invitrogen (CA, USA) |
| | φ80 lacZΔM15 araD139 Δ(ara-leu) 7697 mcrA Δ (mrr-hsdRMS-mcrBC) | |
| *E. coli* BL21 (DE3) | F-ompT gal dcmlonhsdSB(rB-mB-) 1 (DE3 [lacI lacUV5-T7 gene1 ind1 sam7 nin5]) | Novagen (CA, USA) |
| pGEM-T | Ampr; PCR cloning vector | Promega (WI, USA) |
| pUC 19 | Ampr; cloning vector | Stratagene (CA, USA) |
| pET-32a | Ampr; expression vector, T7 promoter | Novagen (CA, USA) |

*PCR*: polymerase chain reaction.

Soil samples are collected in 50-ml sterile polystyrene tube. The collection sites are selected at random including natural habitats such as areas around the oil-fields, river banks and effluents from paper mills, bakery and dairy. From each location a minimum of five samples have to be collected. The collected soil samples are transported to laboratory in sterile plastic bags for storage at 4°C. Total DNA extraction, bacteria isolation and further analysis are carried out from these samples within 7 days.

Each soil sample is air dried, weighed and then physical and chemical characterizations are carried out. For soil particle size analysis, each sample is pretreated with hydrogen peroxide to remove the organic materials and then dispersed using sodium hexametaphosphate and sodium carbonate. Wet sieving is done to separate the soil particles having 0.060 mm or more diameter. The pH of soil is determined in 1:1 (w/w) soil–water slurry. Total carbon (TC), total hydrogen (TH) and total nitrogen (TN) contents are determined using carbon hydrogen nitrogen (CHN) analyser (Perkin Elmer Series II, 2400, USA) using 500 mg of soil sample. Dissolved organic carbon (DOC) contents are determined using a carbon analyser (Elementar, Liqui TOC, Germany). For DOC, water extracts are made by shaking 1.0 g soil into 10 ml distilled water in a horizontal shaker for overnight, followed by filtering through a 45 μm filter paper and acidifying with 50 μl of 10% hydrochloric acid (HCl).

The investigation is carried out to isolate genes responsible for the efficient production of lipase enzyme using metagenomic and traditional genomic approaches. The soil samples are collected from natural habitats such as the surrounding of oil fields, river site, and effluents from paper, bakery and dairy sites of Assam, India.

Soil samples used in the present investigation have exhibited diverse physicochemical properties. The physical and chemical properties of soil used in this study were diverse. The soil is classified as granular and sandy clay on the basis of particle size. Sites used to collect soil samples are presented in Figure 5.1. Data obtained are presented in Tables 5.2, 5.3 and 5.4.

**FIGURE 5.1**   Sampling sites.

**TABLE 5.2**   Physical Characterization of Soil.

| Soil sample | Structure | Texture | Compaction | Amount % of | | |
|---|---|---|---|---|---|---|
| | | | | Sand | Silt | Clay |
| KB1 | Granular | Sandy clay | Moderate | 49 | 11 | 40 |

**TABLE 5.3**   Chemical Characterization of Soil.

| Moisture (%) | pH | Temp | %C | %H | %N |
|---|---|---|---|---|---|
| 21.3 | 6.3 | 28°C | 11.8 | 0.75 | 5.37 |

**TABLE 5.4**   Analysis of Metal Contents in Soil Sample.

| Soil sample | Metal content (µg/gram dry weight soil) | | | | |
|---|---|---|---|---|---|
| KB1 | Ca | Fe | K | Mg | Na |
| | 7117 | 11,978 | 7589 | 3073 | 4642 |

Environment is the vast reservoir of microbes. Torsvik et al. (1996) reported the reassociation kinetics of DNA isolated from various soil samples and the number of distinct prokaryotic genomes has been estimated to be from 2000–18,000 per gram. The complexity of microbial diversity present in soil samples results from multiple interacting physiological parameters such as pH, moisture, soil structure, climatic variations and biotic activity. The metabolism of soil microorganisms is strongly influenced by the availability of water and nutrients. According to Daniel (2005) soil comprises mineral particles of different sizes, shapes and chemical characteristics together with the soil biota and organic matter complexes and the stabilization of clay, sand and silt particles through the formation of aggregates are the dominant structural characteristics of soil matrix. Soil microorganisms often strongly adhere or adsorb into soil particles such as sand grains or clay organic matter complexes. The soil sample KB1 was classified as granular sandy clay soil on the basis of particle size and percentages of carbon, hydrogen and nitrogen were 204, 11.8%, 0.75% and 5.37%, respectively (Table 5.2). The soil sample had moisture content of 21.3%. The elemental content such as Ca, Fe, K, Mg and Na were 7117, 11,978, 7589, 3073 and 4642 µg/g of dry soil, respectively (Table 5.2c). Soil pH and temperature were 6.3 and 28°C respectively. Usually, a very high percentage of soil microorganisms adhere to sandy clay soil. Purdy et al. (1996) reported that high content of clay soil particles tend to absorb organic matter including DNA. According to Jenne et al. (1988), clay is a particle of small size and, thus, has a large surface area per unit weight and generally is coated with metal oxides and organic matter. Clay particles exhibit surface charges that attract both negatively and positively charged ions. Therefore, this study can envision on the role of the different physiological parameters in shaping up the bacterial diversity and complexity present in soil. Data obtained on physicochemical characterization of soil is in agreement with the characters described by Handelsman.

## 5.2 EXTRACTION, PURIFICATION AND QUANTIFICATION OF SOIL METAGENOMIC DNA

The mgDNA is isolated using a modified protocol based on Porteous et al. (1997). The same is presented below:

1. Weighed 750 mg of soil sample in 2 ml microfuge tube.
2. Added 1.0 ml of phosphate buffered saline (PBS) buffer (pH 8) to the soil sample.
3. Vortexed for 5 min and centrifuged at 3000 g for 10 min.

4.  Supernatant is transferred to 2.0 ml microfuge tube and 70 µl of lysis buffer (1.5 M sodium chloride, NaCl), 0.1 M disodium ethylenediaminetetraacetate, Na2EDTA), 4% sodium dodecyl sulphate (SDS) added followed by incubation at 72°C for 45 min.

5.  Microfuged the sample at 13,000 g for 5 min at 4°C and the supernatant transferred to a fresh 2-ml centrifuge tube.

6.  An aliquot of 100 µl of 6 M potassium acetate and 400 µl of 50% polyethylene glycol (PEG) are added to the supernatant and the mixture is allowed to precipitate for 20 min at $-20°C$ and centrifuged at 4°C for 5 min.

7.  The supernatant is removed and the pellet air dried.

8.  The pellet is dissolved in 500 µl TE (TrisHCL EDTA) buffer (pH: 8.0) and then 500 µl of chloroform added followed by centrifugation at 13,000 g at 4°C for 5 min.

9.  The chloroform extraction is repeated twice and 500 µl of isopropanol added to the supernatant and then allowed to precipitate the aqueous DNA for 5 min at 4°C and again centrifuged at 13,000 g for 5 min.

10. The DNA pellet is suspended in 100 µl of 1X TE (10 mm TrisHCl and 1 mm EDTA).

## 5.2.1 PROCEDURE

Agarose gel 0.8% is prepared in 1X TAE (Tris base, acetic acid and EDTA) buffer and heated to dissolve the agarose. It is allowed to cool down around 60°C and then ethidium bromide (EtBr) (10 mg/ml stock) added to make the final concentration of 0.5 g/ml. The gel solution is poured into the gel caster sealed with adhesive tape and fitted with comb. The comb and adhesive tape are removed when gel is solidified. The gel is placed in the electrophoresis chamber filled with 1X TAE buffer, and then DNA samples are loaded in the wells of the gel. The samples are run at 75 V/cm till the loading dye reached 75% of the gel. The gel is removed from the electrophoresis tank and examined on ultraviolet (UV) transilluminator. Metagenomic DNA from the soil samples is isolated using the indirect DNA isolation method following Porteous and Armstrong322 with some modifications. The isolated mgDNA is electrophoresed in 0.8% agarose gel along with 1 kb DNA ladder (MBI Fermentas, Germany) which showed high quality, intact and non-sheared DNA. The size of the extracted mgDNA is found to be greater than 23 kb and the same is shown in Figure 5.2.

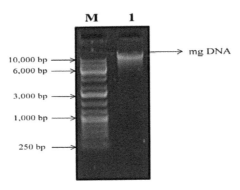

**FIGURE 5.2**   Isolation of metagenomic DNA from soil. Lane M–1 kb DNA ladder (MBI Fermentas, USA); lane 1–mgDNA from soil.

## 5.2.2   SPECTROPHOTOMETRIC ASSESSMENT OF YIELD AND PURITY OF THE MGDNA

Extracted mgDNA is assessed for purity and yield by using the absorbance ratios at 260/230 nm (DNA/humic acids) and at 260/280 nm (DNA/ protein) in UV spectrophotometer, (Thermoscientific, UV-10, Japan). Yield and purity of the isolated mgDNA are presented in Table 5.5.

**TABLE 5.5**   Yield and Purity of mgDNA

| Soil sample | Crude DNA (µg/g soil) | Crude DNA (ng/µl) | Optical density at (A260/A280) | Optical density at (A260/A230) |
|---|---|---|---|---|
| KB1 | 3.8±0.04 | 36±0.2 | 1.6 | 0.8 |

The soil mgDNA was extracted indirectly after subsequent separation of bacterial cells from the matrix followed by the cell lysis and DNA extraction. The protocol involves cell lysis using lysis buffer to disrupt the bacterial cell wall followed by incubation at 72°C for 45 min. According to Rajendhran and Gunasekaran (2008) chemical lysis is a gentle method with less DNA shearing states that indirect method specifically targets prokaryotic DNA, minimizes the extraction of extracellular DNA, and provides larger fragment DNA with a high degree of purity. Though SDS is the most widely used detergent for cell lysis it may not lyse some Gram-positive bacteria; but incubation at 72°C for 45 min in the high salt-SDS buffer produced double amount of DNA as compared to the bead-beating or lysis at 72°C (Gray and Herwig, 1996; Bollet et al., 1991).

Deproteinization with NaCl allows soil particles to precipitate with the cell debris and proteins (Selenska and Klingmüller, 1991). Extraction buffer pH also plays a vital role in the recovery of mgDNA from soil. PEG was used to precipitate the soil mgDNA (Frostegard et al., 1999), it greatly reduces the co-precipitation of humic substances and results almost fourfold reduction in the content of humic substances without decreasing the DNA yield. DNA was precipitated from aqueous phase by the addition of isopropanol. The DNA yield was 3.8 µg/g with the concentration of 36 ng/µl. A high 260/230 ratio (>2) is indicative of pure DNA, whereas a low ratio of humic acid contamination, high 260/280 ratio (>1.7) is indicative of pure DNA, whereas low ratio of protein contamination (Jackson et al., 1997).

## 5.2.3   COMPARISON OF THE DNA EXTRACTION METHODS

The mgDNA from soil sample is isolated using five different methods so as to assess the best one. These methods (M1, M2, M3, M4 and M5) are compared on the basis of processing time, DNA yield, purity and suitability for PCR and restriction digestion. The details of the isolation methods are presents in Table 5.6. M4 is a commercial miniprep kit and is used as per manufacturer's instruction (Mobio Ultraclean soil DNA isolation kit). M5 is a modification of the protocol described by Porteous et al. (1997).

**TABLE 5.6**   Methods Used for the Isolation of mgDNA from Soil Samples.

| Method | Extraction buffer | Cell lysis | Humic acid removal |
|---|---|---|---|
| M1 | EDTA,CTAB, TrisHCl, NaCl, NaPO$_4$ | SDS | CTAB |
| M2 | NaCl, TrisHCl, EDTA | SDS, Vortex | PVPP |
| M3 | EDTA, NaCl, TrisHCl | Bead beating, SDS | PEG |
| M4 | As per Mobio kit | As per Mobio kit | As per Mobio kit |
| M5 | EDTA, SDS, NaCl | Vortex, Heating | PEG |

*EDTA:* ethylenediaminetetraacetic acid, *CTAB:* hexadecyltrimethylammonium bromide, *NaCl:* sodium chloride, *NaPO$_4$:* sodium phosphate, *SDS:* sodium dodecyl sulphate, *PVPP:* polyvinylpolypyrrolidone, *PEG:* polyethylene glycol.

### 5.2.4   METHODS OF PURIFICATION OF CRUDE DNA

The mgDNA extracted from soil samples using M3 and M5 are purified following five different methods:

1. Sephadex G-50 gel filtration (MP1)
2. Silica membrane based spin column purification (Ultraclean, Mobio) (MP2)
3. Silica membrane based spin column purification (Qiagen) (MP3)
4. Agarose gel electrophoresis (electroelution) (MP4)
5. Agarose gel with polyvinylpyrrolidone (PVP) electrophoresis (electroelution) (MP5)

### 5.2.5   PURIFICATION USING SEPHADEX COLUMN

Sephadex G-50 slurry is swollen overnight and packed into spin columns to settle down. Each of the DNA sample (100 μl) to be purified is loaded into the column and kept at room temperature for 5 min and centrifuged at 3000 rpm for 5 min.

### 5.2.6   PURIFICATION USING SILICA MEMBRANE BASED SPIN COLUMN (COMMERCIAL KIT)

The DNA sample is purified using silica membrane based commercial spin column. Each DNA sample (50 μl) is loaded into the column (ultraclean soil DNA isolation kit Mobio, USA). As per manufacturer's instructions the column is kept at room temperature for 5 min and the purified DNA sample eluted in 100 μl of TE Buffer.

### 5.2.7   ELECTROELUTION-BASED PURIFICATION

Each DNA sample (50 μl) is loaded and resolved in 0.8% agarose. High molecular weight band of mgDNA is cut and transferred into a dialysis bag containing three volumes of electrophoresis buffer. The DNA is eluted in to the dialysis bag by electrophoresis for 1.5 h. The DNA sample is precipitated with isopropanol and washed with 70% ethanol followed by air drying. The sample is suspended in 100 μl TE buffer.

## 5.2.8 AGAROSE GEL PURIFICATION

Humic acid co-migrates with nucleic acid under standard electrophoretic conditions. Addition of PVP to agarose gel halts the co-migration of humic compounds by retarding its electrophoretic mobility. Each DNA sample is loaded on 0.8% agarose gel containing 2% of PVP.

## 5.2.9 QUANTIFICATION OF MGDNA AND HUMIC ACID

Quantification of DNA ($A_{260}/A_{280}$) is commonly performed to determine the average DNA concentration and its purity in a solution. Quant-iT Pico-green dsDNA kit (Molecular Probes, USA) is used for the cause as per manufacturer's protocol. Fluorescence is measured using SpectraMax fluorescence microplate reader (Molecular devices, USA) at an excitation of 480 nm and emission of 520 nm. Serially diluted λ phage DNA (1.0–100 ng/ml) is used to prepare the standard curve. The quantification of humic acid is done by absorbance of DNA sample at 340 nm using a spectrophotometer (Thermo Scientific, UV-10, Japan). The concentration of humic acid is calculated based on the standard curve prepared with serial dilution (0.1–100 μg/ml) of commercial humic acid (Merck, India). Humic compounds absorb illumination at 230 nm, protein at 280 nm, whereas DNA at 260 nm. Therefore, the absorbance ratio at 260/230 nm (DNA/humic acid) and 260/280 nm (DNA/protein) are used to evaluate the purity of the soil mgDNA.

The mgDNAs isolated by five different methods (M1–M5) are compared on the basis of yield, purity, processing time, suitability for PCR and digestibility by restriction enzymes. The purity and yield of the isolated mgDNA is calculated by taking optical density (OD) at 230, 260 and 280 nm in the UV-10 spectrophotometer (Thermo Scientific, Japan). The highest DNA yield is obtained in the case of M5, followed by M3. The mgDNA extracted using M5 method has represented both high yield with low concentration of humic acid in comparison to the other methods. Data obtained are presented in Table 5.7 and Figure 5.3.

**TABLE 5.7**  Yield, Purity and Other Useful Parameters of Crude DNA Isolated by Five Different Methods.

| Method | Crude DNA yield (µg/g soil) | Crude DNA conc. ng/µl | Humic acid conc. ng/µl | OD at A260/ A280 | OD at A260/ A230 | Processing time (h) | 16S rDNA amplification | Endonuclease activity |||
|---|---|---|---|---|---|---|---|---|---|---|
| | | | | | | | | *EcoRI* | *BamHI* | *HindIII* |
| M1 | 2.8±0.02 | 3.0±0.04 | 72±3.2 | 1.3 | 0.5 | 5.0 | − | − | − | − |
| M2 | 3.2±0.08 | 6.8±0.14 | 86±0.2 | 1.2 | 1.2 | 3.5 | − | − | − | − |
| M3 | 3.8±0.02 | 26±0.08 | 62±3.0 | 1.5 | 0.6 | 7.0 | − | − | − | − |
| M4 | 0.6±0.06 | 3.0±0.12 | 4.0±0.02 | 1.6 | 0.9 | 1.5 | − | + | + | + |
| M5 | 3.8±0.04 | 36±0.2 | 8.0±0.13 | 1.6 | 0.8 | 2.5 | − | + | + | + |

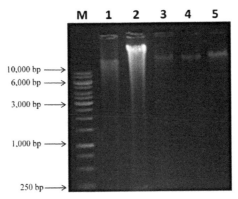

**FIGURE 5.3**  Metagenomic DNA extracted from soil sample using different methods. Lane M–1 Kb DNA ladder (MBI Fermentas, USA); lane 1–5 metagenomic DNA extracted using M1–M5 respectively.

The modified M5 method of soil mgDNA extraction yielded 3.8±0.04 µg DNA per g soil having the DNA concentration 36±0.2 ng/µl. The method reduced humic acid concentration to 8.0±0.13 ng/µl which is the primary goal of the present investigation. Majority of the DNA extraction methods from soil samples caused high humic acid contamination affecting the PCR and restriction endonuclease digestion. The M5 method could overcome the hindrance caused by humic acid contamination with the reduction of 85–90% of humic acid. The method was found to be the best for the extraction of mgDNA from soil samples with lesser humic acid contamination and higher yield of quality DNA.

## 5.3 ISOLATION AND PURIFICATION OF MGDNA FROM GRAM⁺ AND GRAM⁻ BACTERIA

Method M5 is used to isolate genomic DNA from *Bacillus subtilis* and *E. coli* as representatives of Gram-positive and Gram-negative DNA, respectively to validate the utility of the extraction method for cultivable bacteria.

Genomic DNA is isolated from Gram-positive and Gram-negative bacteria using the method M5 to confirm the suitability of this method in the case of cultivable bacteria which is shown in Figure 5.4. It is observed that the method gives higher concentration of DNA in the case of Gram-positive and Gram-negative bacteria with 58.85 and 73.48 ng/μl in *B. subtilis* (MTCC 121) and *E. coli* MTCC 40), respectively. Yield and purity of the isolated DNA are presented in Table 5.8.

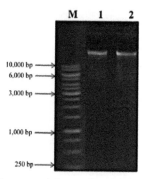

**FIGURE 5.4**  Genomic DNA extracted from culturable bacteria using M5 methods. Lane M–1 kb DNA ladder (MBI Fermentas, Germany); lane 1–genomic DNA extracted from *E. coli* (MTCC 40) Gram-negative bacteria; lane 2–genomic DNA extracted from *B. subtilis* (MTCC 121) Gram-positive bacteria.

**TABLE 5.8**  Yield and Purity of Genomic DNA Isolated from Gram Positive and Negative Bacteria Using M5 Method.

| Sample | DNA yield μg/μl | DNA conc. ng/μl | Optical density at A260/A280 |
|---|---|---|---|
| *B. subtilis* (MTCC 121) | 0.14 | 58.85 | 1.58 |
| *E. coli* (MTCC 40) | 0.44 | 73.48 | 1.71 |

All extracted soil mgDNA samples are used for 16S rDNA-based amplification and restriction digestion to determine the quality but none found to be suitable for PCR and/or restriction digestion. The low quality DNA extracted using M4 method (commercial kit) is digested separately with *EcoR*I, *Bam*HI and *Hind*III, but none of them was found to be suitable for PCR. Therefore, the DNA extracted by using this commercial kit required further purification for the downstream application of mgDNA library construction.

DNAs extracted using the methods M3 and M5 are subjected to further purification. However, DNA extracted using the methods M4 and M5 is suitable for digestion by all the 3 restriction enzymes but negative for PCR. Therefore, purification of DNA by these two methods is evaluated using the purification strategies of commercial spin column, gel filtration and electroelution. All these five mgDNA extraction and purification methods are evaluated to confirm the suitability of the extracted DNA for the subsequent molecular analyses. The PCR amplification of 16S rDNA from mDNA after purification by five different methods is shown in Figure 5.5.

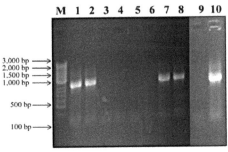

**FIGURE 5.5** PCR amplification of 16S rDNA from mgDNA after purification by five different methods. Lane M–1 Kb DNA ladder (Gene Ruler, USA); lane 1–2—after purification by MP1 of DNA extracted by M3 and M5; lane 3–4—after purification by MP2 of DNA extracted by M3 and M5; lane 5–6—after purification by MP3 of DNA extracted by M3 and M5; lane 7–8—after purification by MP4 of DNA extracted by M3 and M5; lane 9–10—after purification by MP5 of DNA extracted by M3 and M5 respectively.

The method MP1 yielded pure DNA suitable for both PCR and restriction digestion, and DNA recovery is also high in Sephadex (G-50) column-based purification. DNA recovery before purification is considered as 100%. The recovery of DNA in the use of method M5 using MP1 has showed 98% recovery. The endonuclease activity of purified mgDNA extracted using methods M3 and M5 are shown in Figures 5.6 and 5.7.

The data obtained on DNA recovery, 16S rDNA amplification and endo-nuclease activity after purification using different method are presented in Table 5.9.

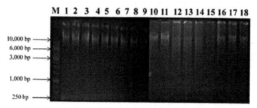

**FIGURE 5.6**   Restriction digestion of mgDNA extracted using M3 method with *Eco*RI, *Bam*HI and *Hind*III before and after purification by five different methods. Lane M–1 Kb DNA ladder, MBI before purification; lane 4–6—restriction digestion with *Eco*RI, *Bam*HI and *Hind*III after MP1 purification; lane 7–9—restriction digestion with *Eco*RI, *Bam*HI and *Hind*III after MP2 purification; lane 10–12—restriction digestion with *Eco*RI, *Bam*HI and HindIII after MP3 purification; lane 13–15—restriction digestion with *Eco*RI, *Bam*HI and HindIII after MP4 purification; lane 16–18—restriction digestion with EcoRI, BamHI and HindIII respectively after MP5 purification.

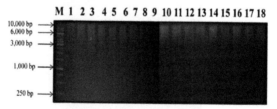

**FIGURE 5.7**   Restriction digestion of mgDNA extracted using M5 method with *Eco*RI, *Bam*HI and *Hind*III before and after purification by five different methods. Lane M–ladder, MBI Fermentas, USA; lane 1–3—restriction digestion with *Eco*RI, *Bam*HI and *Hind*III before purification; lane 4–6-restriction digestion with *Eco*RI, *Bam*HI and *Hind*III after MP1 method purification; lane 7–9 restriction digestion with *Eco*RI, *Bam*HI and *Hind*III after MP2 purification; lane 10–12-restriction digestion with *Eco*RI, *Bam*HI and *Hind*III after MP3 purification; lane 13–15- restriction digestion with *Eco*RI, *Bam*HI and *Hind*III after MP4 purification; lane 16–18—restriction digestion with *Eco*RI, *Bam*HI and *Hind*III respectively after MP5 purification.

**Table 5.9** DNA Recovery, Suitability for PCR and Restriction Digestion After Purification by Five Different Methods.

| Method | DNA recovery (%) | | 16S rDNA PCR | | Endonuclease activity | | | | | |
|--------|------|------|------|------|------|------|------|------|------|------|
| | | | | | *Eco*RI | | *Bam*HI | | *Hind*III | |
| | M3 | M5 | M3 | M5 | M3 | M5 | M3 | M5 | M3 | M5 |
| MP1 | 96 | 98 | + | + | + | + | + | + | + | + |
| MP2 | 80 | 78 | – | – | – | + | – | + | – | + |
| MP3 | 81 | 86 | – | – | – | + | – | + | – | + |
| MP4 | 40 | 41 | + | + | – | + | + | + | + | + |
| MP5 | 13 | 20 | – | + | – | + | – | + | – | + |

## 5.4   POLYMERASE CHAIN REACTION

Metagenomic DNA amplification is performed with GeneAmp PCR system (Applied Biosystem, USA). PCR reaction mixture contained 1X PCR buffer, 200 μM of each deoxynucleotides (dNTPs), 3.0 μM magnesium chloride (MgCl$_2$), 0.2 μM of each forward and reverse primers and 2.5 U of Taq DNA polymerase (SIGMA, USA) in 50 μl reaction volume. The PCR product is sequenced bidirectionally using both forward and reverse primers of the 16S rDNA. 16S rRNA gene in the mgDNA is amplified using the universal primers to confirm the suitability of mgDNA for PCR, restriction digestion and cloning experiments. The forward primer B 27F (5′ AGA GTT TGA TCC TGG CTC AG 3′), and the reverse primer U 1492R (5′ GGT TAC CTT GTT ACG ACT T 3′) are used for PCR amplification. For the positive control 5 ng of *E. coli* (MTCC40) or *B. subtilis* (MTCC121) genomic DNA is used; and a sample without template as the negative control. PCR is performed by subjecting a reaction mixture to initial denaturation at 94°C for 2 min, followed by 35 cycles of 94°C for 45 s, annealing at 55°C for 1 min and extension at 72°C for 2 min followed by the final extension at 72°C for 10 min in a thermal cycler (Applied Biosystem). The amplification is determined by electrophoresis of reaction product in 1% agarose gel.

## 5.5   RESTRICTION DIGESTION BY *Eco*RI, *Hind*III AND *Bam*HI

To examine the suitability of mgDNA for restriction digestion, 0.25 μg of each DNA sample is digested separately with 2.5 units of *Eco*RI, *Hind*III and *Bam*HI (MBI Fermentas, Germany) restriction enzymes in a 25 μl reaction mixture. The mixtures are incubated at 37°C for 4 h followed by inactivation of the restriction enzymes by heating at 70°C for 10 min. The digested products are resolved on 0.8% agarose gel.

## 5.6  METAGENOMIC DNA LIBRARY CONSTRUCTION

Soil mgDNA library is constructed using pUC-19 as the cloning vector. Purified mgDNA is partially digested using *Bam*HI (MBI Fermentas, Germany) restriction enzyme. The digested product is resolved in 0.8% agarose gel and DNA fragments ranging about 0.5–2.0 kb are fractionated by agarose gel purification using QIAquick gel extraction kit (Qiagen, Germany). The purified mgDNA fragments ranging about ~0.5–2.0 kb are separated and ligated to *Bam*HI-digested and dephosphorylated pUC19 cloning vector using T4 DNA ligase (MBI Fermentas, Germany) at 22°C overnight. The ligated mixture is transferred to *E. coli* DH10B by electroporation (200 Ω, 25 µF and 2.5 kV) using Gene Pulser (Biorad, USA). The undigested pUC-19 cloning vector is transferred to DH10B competent cells as the positive control to confirm the transformation efficiency. Transformed cells are cultured on Luria broth (LB) agar plates supplemented with ampicillin (100 µg/ml) and X-gal (5-bromo-4-chloro-3-indolyl-β-D-galactopyranoside) (20 µg/ml). The recombinants are scored by blue–white screening after overnight incubation at 37°C. The resulted library is stored in 15% glycerol at −80°C.

## 5.7  SEQUENCE-BASED METAGENOMIC APPROACH

The amplification is carried out as described with GenAmp PCR system (Applied Biosystem, USA). The optimal PCR reaction condition for 16SrDNA amplification of soil metagenome is presented:

| Conditions | Time frame/cycles |
|---|---|
| Initial denaturation | 94°C for 5 min |
| Denaturation | 94°C for 45 s |
| Annealing | 55°C for 1 min |
| Elongation | 72°C for 2 min |
| Final elongation | 72°C for 10 min |
| Cycles | 35 cycles |

The sequences of the universal primers and PCR condition are:

| Primer name | Primer sequence (5'-3') |
| --- | --- |
| 27F | AGAGTTTGATCMTGGCTCAG |
| 1492R | TACGGYTACCTTGTTACGACTT |

## 5.7.1 ANALYSIS OF PCR PRODUCTS BY AGAROSE GEL ELECTROPHORESIS

The PCR products are separated on 1% (w/v) agarose gel with ethidium bromide (0.5 µg/ml) and 1X TAE (Tris-acetate-ethylenediaminetetraacetic acid (EDTA) running buffer. DNA molecular marker 100-bp ladder (Bangalore genei, India) is used for the evaluation and determination of amplicon size. The amplicon bands are visualized directly upon UV transilluminator at 302 nm. Documentation of the DNA resolved gels is performed using photographic image obtained with a Bio-Rad Gel Documentation system.

## 5.7.2 PURIFICATION OF PCR PRODUCT BY USING QIAquick PCR PURIFICATION KIT

PCR products are purified from primers, polymerases, unincorporated nucleotides, buffer components and salts using the QIAquick PCR purification kit (Qiagen, Germany). Five volume of PB buffer is added to the PCR product. The mixture is shaken and added to the QIAquick spin column and centrifuged at 13,000 rpm for 1 min. The collected flow-through is discarded and DNA is washed with 750 µl PE buffer. PE buffer is removed by 1 min centrifugation at 13,000 rpm followed by 1 min centrifugation to remove residual PE buffer completely from the column. To elute the PCR product from QIAquick spin column, 40 µl sterile water (preheated to 65°C) is added and after incubation for 2 min at room temperature, product is eluted by centrifugation for 1 min at 13,000 rpm.

## 5.7.3 GEL PURIFICATION OF DNA BY QIAQUICK GEL EXTRACTION KIT

PCR product is run on 1% agarose gel in 1XTAE buffer. After separation by electrophoresis, the gel is examined under UV transilluminator and gel slices containing DNA of interest are excised using a sterile scalpel. Excised gel is mixed with three volume of QIAquick gel (QG) buffer, incubated at 50°C for 10 min (or until the gel slice has completely dissolved). To bind DNA, the sample is applied to QIAquick column, and centrifuged for 1.5 min at 13,000 rpm. The flow-through is discarded and the column is placed back in the same collection tube. An aliquot 500 μl QG buffer is added to the column and centrifuged for 1.5 min. Now, 750 μl of PE buffer is added to the column for washing and centrifuged for 1.5 min. The flow-through is discarded and column centrifuged for an additional 1 min at 13,000 rpm to remove the residual buffer. The QIAquick column is placed into a clean 1.5 ml microfuge tube. To elute DNA, 40 μL deionized water (preheated at 65°C) is added to the centre of the QIAquick membrane. The column is allowed to stand for 1 min and centrifuged for 1.5 min. The concentration of the recovered DNA is measured by ND-1000 spectrophotometer (Nanodrop technologies, USA) and used in further experiments.

DNA extraction from Gram-positive bacteria with SDS-based method followed by incubation at high temperature yielded 2–6 fold DNA yield. The genomic DNA was isolated from Gram-positive and -negative bacteria using M5 method. The M5 method promised higher concentration of quality DNA in the case of both Gram-positive and -negative bacteria.

All mgDNA samples were used for 16S rDNA-based amplification and restriction digestion to determine the quality but none was found to be suitable. According to Riesenfeld et al. (2004) PCR and restriction digestion involve successive enzymatic reaction and the enzymes require contamination-free sites. The DNA sample extracted using M4 method (commercial kit) was effectively digested with EcoRI, BamHI and HindIII, but none was suitable for PCR. Hence, the DNA extracted using the commercial kit was further purified. hexadecyltrimethylammonium bromide (CTAB) and PEG addition to the soil buffer slurry before cell lysis minimized co-precipitation of humic substances with DNA. The same was also used by Zhou et al. (1996) and mgDNA isolated using PEG in M3 method was found to be suitable for PCR and restriction; the resultant mgDNA from M4 and M5 methods became suitable for restriction digestion but not for PCR. Therefore, the isolated mgDNAs by M4 and M5 were subjected to further purification with the use of commercial spin column, gel filtration and electroelution. Metagenomic DNA purification using Sephadex G-50 (MP1) was found to be efficient in removing humic substances with minimal loss of DNA and was also suitable for

PCR; the same was reported by Dijkmans et al. (1993). The protocol with the use of silica gel, developed by Schneegurt et al. (2003), was found inefficient in the present investigation. According to Harry et al. (1999) humic substances might compete with DNA for binding sites during purification using the commercial purification columns. Even the commercial gel filtration resins failed to revive the inhibitors that came along with soil DNA. Jackson et al. (1997) and Miller et al. (1999) reported superior separation of DNA from humic substances with the use of Sepharose resins. Tsai and Olson (1991) observed that polyacrylamide and dextran gel separated DNA on the basis of size and could not remove humic substances from the complex. Hilger and Myrold (1991) reported agarose gel electrophoresis to be more efficient in separating DNA from the humic substances. Jackson et al. (1997) reported a comparison of effectiveness of mDNA isolation using Sepharose 4B, Sephadex G-200 and Sephadex G-50 in the case of different soil types and found Sepharose 4B to be more effective in providing good separation. Therefore, it could be stated that the purification of mgDNA does not depend only on the type of purification strategy used but also on the type of humic substances present in the soil sample.

## 5.8   CONSTRUCTION OF 16S RDNA GENE LIBRARIES

In PCR reaction, Taq polymerase enzyme adds a single 3'-A overhang to both ends of the PCR product. The structure of these PCR products favour direct cloning into a linearized cloning vector with single 3'-dT overhang. Such an overhang at the vector cloning site not only facilitates cloning, but also prevents the recircularization of the vector. The PCR products are mixed with pGEM-T Easy (Promega, USA) vector DNA (Fig. 5.8) in 3:1 ratio and ligated at 22°C using T4 DNA ligase. The setting of ligation reactions is described below:

| Contents | Composition (10 µl) |
| --- | --- |
| 2X Rapid ligation buffer | 5 µl |
| pGEM-T Easy vector (50 ng) | 1 µl |
| PCR product | 3 µl |
| T4 DNA ligase (3 weiss units/µl) | 1 µl |

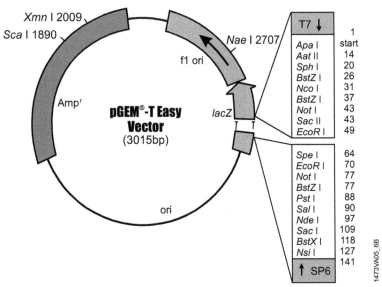

**FIGURE 5.8**   pGEM-T Easy vector map (PROMEGA™).

Reaction mixing is done by pipetting; the reactions are incubated for 1 h at room temperature. In order to obtain the maximum number of transformants, each reaction is incubated overnight at 4°C.

A small mgDNA library is constructed successfully using the cloning vector pUC-19. Further, the recombinants are selected on the basis of blue–white screening after overnight incubation at 37°C temperature. Therefore, mgDNA library is constructed successfully using DNA extracted by the method M5 and purified by MP1.

Among the various culture-independent techniques, 16S rRNA-based method is a widely used technique to describe the composition of complex microbial communities. Polymerase chain reaction for mgDNA amplification is used as an indicator of DNA purity for further downstream processes viz. cloning, transformation and metagenomic library construction.

The mgDNA isolated from the soil samples is amplified using the universal 27F and 1492R primers. After the PCR, the size of amplicon is found to be approximately 1.5 kb in 1% agarose gel electrophoresis.

The purified mgDNA using M5 method was used for the construction of genomic library. The mgDNA library (Fig. 5.9) was constructed using the pUC-19 plasmid vector. The same type of library from soil was constructed by Manjula et al. (2011).

**FIGURE 5.9**   Metagenomic library construction and blue–white screening.

## 5.9   PREPARATION OF *E. coli* COMPETENT CELLS AND TRANSFORMATION

Only competent bacterial cells are bacterial cells that are capable of intaking foreign DNA. The bacterial populations treated with $CaCl_2$ induce a transient state of competence in which the pores are created in the cell wall of bacteria enabling them to uptake DNA. The transient state is maintained at storing the cells at $-70°C$.

### 5.9.1   PROTOCOL

- Volume of 0.5 ml desired *E. coli* (DH5α) overnight bacterial culture is inoculated aseptically into 50 ml of LB antibiotic free medium.
- The liquid bacterial culture is allowed to grow up to early log phase at 37°C in a shaker at ~200 rpm until OD600 reaches 0.6–0.8 (3–3.5 h).
- Bacterial cell suspension is transferred to 50 ml sterile ice-cold Oakridge tube and kept in ice for 20 min.
- Tube is centrifuged at 3500 rpm for 5 min at 4°C.
- Supernatant is discarded and the pellet is resuspended in 9 ml of sterile ice-cold 100 mM $MgCl_2$ solution and mixed gently to suspend the cells completely in the solution.
- Tubes containing resuspended bacterial cells are centrifuged at 3500 rpm for 5 min at 4°C.
- Bacterial cells pellet is resuspended in 9 ml of sterile ice-cold 100 mM $CaCl_2$ solution and mixed gently to suspend the cells completely in the solution.
- Tubes are rested in ice for 40 min.

- Tubes are centrifuged at 3500 rpm for 5 min at 4°C. The pellet obtained is resuspended carefully in 3.5 ml of sterile ice-cold FT buffer (85 mM $CaCl_2$ and 15% glycerol).
- Aliquots of 100 μl resuspended bacterial cells are pipetted to prechilled 1.5 ml Eppendorf tubes using the end cut tips.
- Tubes are kept frozen on dry ice for a few minutes and stored at $-70$ to $-80°C$ until further use.

Transformation is the process by which the recombinant DNA (rDNA) enters into the host cells and then proliferates. It is called as "transformation" because the function of *E. coli* cells is protected by the antibiotic-resistance gene whose product can inactivate the specific antibiotic. This enables selection of the cells having incorporation of rDNA molecule. The transferred rDNA also multiplies along with the host cells to produce as many identical copies of the clones.

### 5.9.2  PROTOCOL

- Vials containing competent bacterial cells are thawed on ice. The ligated product is added to the vial content, mixed gently and kept on ice for 40 min.
- Iced tubes containing competent bacterial cells are quickly removed from ice and heat shocked at 42°C for 90 s without shaking and then placed back on ice for 5 min.
- Aliquot of 250 μl LB broth is added to each of tubes in sterile condition (under a laminar hood) and then inverted twice to mix the contents.
- Tube is incubated at 37°C for 1 h with shaking.
- Aliquots of 100 μl the transformed cells is plated on LBA containing ampicillin (100 μg/ml) plates overlaid with isopropyl β-D-1-thiogalactopyranoside ((IPTG) 6 μl) and X-gal (12 μl) and incubated overnight at 37°C.
- Recombinant clones are selected based on the blue–white screening.

### 5.10  ANALYSIS OF TRANSFORMATION EFFICIENCY USING NON-RECOMBINANT PLASMID

Usually cloning vectors encode a shortened derivative of *lacZ*, which encodes the N-terminal alpha-peptide of beta-galactosidase. *E. coli* strains are capable to produce the C-terminal portion of the enzyme and transformants can use the reconstituted enzyme to break down X-gal. The polylinker site in such vectors is located within the *lacZ* gene. Incorporating an insert into this site will disrupt the gene and recombinants will be unable to break down X-gal. Colonies carrying plasmid with no insert will remain to be blue whereas colonies carrying the recombinant plasmid will be white.

## 5.10.1   PROTOCOL

- Aliquot of 100 μl of the transformed cells is overlaid on the plates of containing LB + ampicillin (100 μg/ml) overlaid with 6 μl of 200 mM IPTG and 12 μl of 100 mg/ml X-gal and then incubated overnight at 37°C.
- White colonies are selected as true recombinants having the insert and blue as transformants without the gene of interest.
- Grids are prepared on LB + ampicillin + IPTG + X-Gal plates and 2–3 colonies along with some white colonies transferred into the grid. The plates are then incubated overnight at 37°C.
- Recombinants are confirmed following blue–white screening.

## 5.11   SCREENING OF RECOMBINANTS BY α COMPLEMENTATION

Blue–white screening is a technique that allows rapid and convenient detection of recombinants in the vector-based cloning experiments. Aliquots of 100 μl transformed cells were plated on LBA + ampicillin (100 μg/ml), overlaid with 6 μL of 200 mM IPTG and 12 μL of 100 mg/ml X-gal and incubated at 37°C. Grids were prepared on LBA + ampicillin + IPTG + X-gal plates and 2–3 blue colonies along with 15 white colonies were transferred into the grid. Plates were incubated overnight at 37°C.

White colonies three appeared within the blue, and on subculture of blue colonies on LBA + ampicillin + IPTG + X-gal remained to be blue. White colonies represented the recombinants whereas blue colonies nonrecombinants and the same are shown in Figures 5.10(a), (b). All three white colonies and two blue colonies further analysed by colony PCR to confirm recombinants and nonrecombinants.

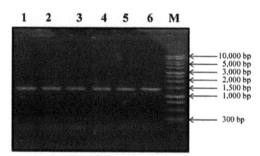

**FIGURE 5.10(a)** 16S rDNA purified product using QIA quick PCR purification kit. Lane M–1kb DNA ladder; lane 1–6—16S rDNA purified product using QIAquick PCR purification kit.

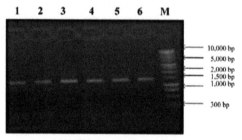

**FIGURE 5.10(b)** 16S rDNA purified product using QIAquick gel extraction kit. Lane M–1kb DNA ladder; lane 1–6—16S rDNA PCR amplified product.

## 5.12 PURIFICATION OF PCR PRODUCTS

After PCR amplification, the triplicate PCR reaction on soil mgDNA samples are pooled and purified to remove the residual enzymes, dNTPs, primers and dimers. The amplified products are purified using QIAquick PCR purification kit and QIAquick gel extraction kit for further experimental analysis (Fig. 5.11).

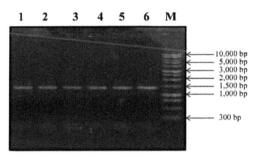

**FIGURE 5.11**   16S rDNA amplification of soil mg DNA. Lane M—300 bp DNA ladder; lane 1–6—16S rDNA PCR amplified products.

Metagenomic clone libraries enable the exploitation of phylogenetic and metabolic diversity of microbes in the environment without their culture. The pGEM-T Easy vector is linearized vector with a single 3′-terminal thymidine at both the ends. Approximately 50 ng of purified PCR products are ligated into pGEM-T Easy vector and incubated over-night at 4°C. The ligated mixture is transferred to *E. coli* (DH5α) competent cells using heat shock method at 42°C, for 90 min.

## 5.13   TRANSFORMANT CONFIRMATION BY COLONY PCR USING UNIVERSAL PRIMERS

Colony PCR is a rapid strategy to screen *E. coli* colonies for the presence of recombinant plasmids. It is a complementary and confirmatory tool for the analysis of recombinant clones by other method such as insertional inactivation and blue–white screening. Colony PCR can be used to determine the insert size and orientation of the gene in the vector. Sequences of the universal primers are as follows:

| Primer name | Primer sequence (5′-3′) |
|---|---|
| M13F | GTTGTAAAACGACGGCCAGT |
| M13R | CACAGGAAACAGCTATGACC |
| 27F | AGAGTTTGATCMTGGCTCAG |
| 1492R | TACGGYTACCTTGTTACGACTT |

The presence of positive clones is confirmed by colony PCR using vector-specific M13 forward, reverse primers and 16S rRNA gene-specific

27F, 1492R universal primers. The amplification of DNA is done as per the procedure described elsewhere. The PCR conditions are presented below:

| PCR conditions | Time frame/cycles |
| --- | --- |
| Initial denaturation | 94°C for 2 min |
| Denaturation | 94°C for 45 s |
| Annealing | 55°C for 1 min |
| Elongation | 72°C for 2 min |
| Final elongation | 72°C for 10 min |
| Cycles | 30 cycles |

Following the composition of amplifications, the amplified DNA is electrophoresed in 1% agarose gel and visualized under UV transilluminator. The amplified product corresponding to the expected size is used for sequencing. All the positive transformants are inoculated in 5 ml LB broth with 100 µg/ml of ampicillin. The tubes are cultured overnight shaking at 200 rpm at 37°C for subsequent isolation of the recombinant plasmid DNA.

### 5.13.1 POSITIVE TRANSFORMANTS BY USING COLONY PCR

The white colonies (3) and blue colonies (2) were randomly picked and streaked on replica plates. Recombinant and nonrecombinant plasmids were confirmed using the colony PCR vector-specific universal M13 F and M13 R primers. The amplicons were analysed on 1.2% (w/v) agarose gel. The size of insert was found to be approximately 1.5 kb (Fig. 5.12).

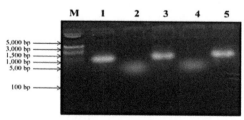

**FIGURE 5.12**   Amplification of insert by colony PCR. Lane M–100 bp Gene Ruler DNA ladder; lane 1, 3, 5—PCR amplified product of white colonies; lane 2, 4–PCR amplified product of blue colonies.

All recombinant plasmids were further checked for the presence of the insert and size of the insert by PCR using 16S rRNA gene specific 27F and 1492R universal primers. The PCR products were analysed on 1.2% (w/v) agarose gel (Fig. 5.13). PCR of recombinant plasmids isolated from white colonies; lane M—100 bp DNA ladder; lane 1–3—PCR amplified product of plasmid isolated from white colonies.

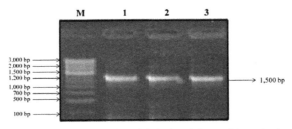

**FIGURE 5.13**  PCR of recombinant plasmids isolated from white colonies. Lane M–100 bp DNA ladder; lane 1 to 3—PCR amplified product of plasmid isolated from white colonies.

## 5.13.2   ISOLATION OF PLASMID DNA FROM TRANSFORMED E. coli

The recombinant plasmids are isolated from the overnight culture of transformed bacterial cells using alkali lysis method (Birnboim and Doly, 1979) with minor modifications.

## 5.13.3   REAGENTS

**Solution I:** Glucose 50 mM, TrisCl (pH 8.0) 25 mM and EDTA (pH 8.0) 10 mM
**Solution II (freshly prepared):** NaOH 0.2 N and SDS 1%
**Solution III:** Potassium acetate 5 M, glacial acetic acid 11.5 ml and distilled water 28.5 ml

### 5.13.4   PROTOCOL

- Overnight culture-derived cells are pelleted by centrifuging 4 ml of fresh bacterial culture supplemented with 100 µg/ml ampicillin at 10,000 rpm for 5 min in a refrigerated Beckman Centrifuge, UK and supernatant was decanted.
- Cell pellet is suspended in 100 µl of ice-cold solution I with vigorous vortexing.
- Aliquot of 200 µl of freshly prepared alkaline-lysis solution II is added to each of the bacterial suspensions.
- Contents are mixed by inverting the tube gently and incubated on ice for 5 min.
- The 150 µl of ice-cold alkaline lysis solution III is added to each tube and dispersed through the viscous bacterial lysate by inverting the tube gently 5–6 times and incubated on ice for 15 min.
- Bacterial lysate is centrifuged at 13,000 rpm for 10 min at 4°C.
- Supernatant is transferred to the fresh tube and an equal volume of ethanol is added at room temperature to precipitate the DNA. The mixture is allowed to stand for 2 min at room temperature.
- Tube is centrifuged at 10,000 rpm for 5 min at 4°C. Following the removal of supernatant, the pellet is washed with 500 µl of 70% ethanol.
- Plasmid DNA is recovered by centrifugation at 13,000 rpm for 5 min at 4°C.
- Following carefully draining the supernatant, each tube is kept open at room temperature until the ethanol gets evaporated and no fluid remained visible in the tube.
- Pellet is dissolved in 50 µl of TE and stored at $-70°C$.

### 5.14   ISOLATION AND RESTRICTION OF RECOMBINANT/ NONRECOMBINANT PLASMID DNAS

To confirm the presence of the gene insert in the recombinant plasmid, recombinant and nonrecombinant plasmid DNAs are digested with *Eco*RI restriction enzyme. The reaction components are mixed gently and incubated at 37°C in a water bath for 3 h. After restriction digestion, all samples are subjected to 1% (w/v) agarose gel electrophoresis. The reaction mixture is described below:

| Components | Volume (µl) |
|---|---|
| Plasmid DNA | 15 |
| 10x buffer | 2 |
| *EcoR*I (20 U/µl) | 2 |
| Distilled water (nuclease free) | 1 |
| Total volume | 20 |

### 5.14.1 PLASMID FROM RECOMBINANTS

All white colonies having the insert of 1.5 kb were inoculated separately in LB medium supplemented with ampicillin (100 µg/ml) for plasmid isolation. The isolated plasmids were of high quality with the purity of A260/A280 1.76–1.94 and concentration 400–2300 ng/µl. The quality was analysed on 1% (w/v) agarose gel.

Plasmids were digested using *EcoR*I restriction enzyme to confirm the presence of the insert. After the restriction digestion the digested products were analysed on 1% (w/v) agarose gel. After restriction digestion of the recombinant plasmid showed two bands while nonrecombinant plasmid showed single band. Size of the insert was found to be 1.5 kb (Fig. 5.14).

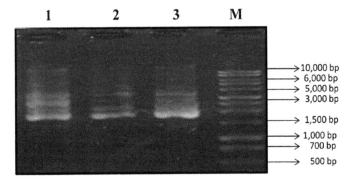

**FIGURE 5.14**   Isolation of recombinant plasmid from white colonies. Lane M–1 kb DNA ladder; lane 1–3—plasmid isolated from white colonies.

## 5.15   ANALYSIS OF 16S RDNA GENE

### 5.15.1   SEQUENCING OF RECOMBINANT PLASMIDS AND ANALYSIS OF 16S rDNA LIBRARY

The isolated plasmids from all the positive transformants are sequenced using vector specific M13 forward and reverse primers. The recombinant plasmids are sequenced with Big Dye Terminator version 3.1 cycle sequencing kit and ABI 3500XL Genetic Analyser. The composition of the sequencing reaction mixture is given below:

| Components | Volume (µl) |
|---|---|
| Big Dye Terminator Ready Reaction Mix | 4.0 |
| Template (100 ng·µl-1) | 1.0 |
| Primer (10 pmol·λ-1) | 2.0 |
| Milli Q water | 3.0 |
| Total volume | 10.0 |

### 5.15.2   PCR CONDITIONS FOR SEQUENCING

The PCR conditions for DNA sequencing are given below:

| Condition | Time frame |
|---|---|
| Initial denaturation at 96°C | 1 min |
| Denaturation at 96°C | 10 s |
| Hybridization at 50°C | 5 s |
| Elongation at 60°C | 4 min |

Sequences of the vector DNA are searched using the online programme VecScreen (http://www.ncbi.nlm.nih.gov/VecScreen/VecScreen.html) and trimmed manually. The overlapping region between the forward and reverse sequences is searched, aligned and assembled together using BioEdit editor. The clone sequences are chimera checked using Bellerophon. All validated sequences are analysed for similarity using the online BLASTn algorithm of NCBI (http://ncbi.nlm.nih.gov/blast/blast.cgi).

The gene sequences obtained from the clone libraries are compared with existing closely matched gene sequences in the databases for the estimation of similarity and evolutionary distance of the cloned sequences.

## 5.16 PHYLOGENETIC ANALYSIS

The 16S rDNA sequence of bacteria under study is aligned with reference sequences showing homology with the NCBI database using the multiple sequence alignment programme of MEGA 4.0. Phylogenetic tree is constructed using ClustalW by distance matrix analysis and neighbour joining method. Phylogenetic trees are displayed using TREEVIEW.

## 5.17 GENBANK SUBMISSION AND ACCESSION NUMBERS

The 16S rDNA gene sequences determined in the investigation is deposited in the GenBank of NCBI data library (http://www.ncbi.nlm.nih.gov/GenBank) under different accession numbers with respect to each sequence. The 16S rRNA-cloned gene sequences were deposited in the GenBank and accession number of the partial 16S-rDNA sequences was given to be KJ685805, KJ700876 and KJ700877for KBSR1, KBSR2 and KBSR3, respectively.

All three 16S rRNA cloned genes were sequenced using M13F and M13R primers and named KBSR1, KBSR2 and KBSR1. The clones KBSR1, KBSR2 and KBSR3 possessed the sizes of 687, 986 and 700 bp, respectively.

## 5.18 PHYLOGENY

### 5.18.1 PHYLOGENY OF CULTURABLE BACTERIA: KBSR1

The homologous search result of the culturable bacteria KBSR1 demonstrated 97% similarity of 16S rDNA sequence with the uncultured bacterium gene for 16S rRNA, partial sequence, clone:0502TCLN280. The phylogenetic tree was constructed from the sequence data by the Neighbour-Joining method which showed the uncultured bacterium (GenBank FR687433) with 97% homology with KBSR1 and represented its closest phylogenetic neighbour. The phylogenetic tree of KBSR1 is shown in Figure 5.15.

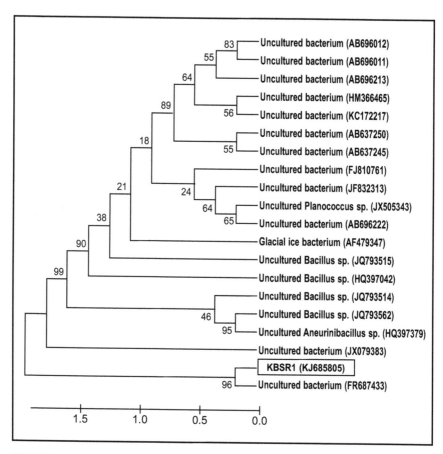

**FIGURE 5.15** Phylogenetic relationship of KBSR1 and other closely related uncultured bacterium based on 16S rDNA sequencing. The tree was generated using the neighbour joining method. The data set was re-sampled 1000 times by using the bootstrap option and percentage values are given at the nodes.

The homologous search result of KBSR2 demonstrated 99% similarity of 16S rDNA sequence with the uncultured *Enterobacter* sp. clone czt7 16S ribosomal RNA gene, partial sequence. The phylogenetic tree was constructed from the sequence data by the neighbour joining method which showed the uncultured bacterium (GenBank JN990086) with 99% homology with KBSR2 and represented its closest phylogenetic neighbour. The phylogenetic tree of KBSR2 is shown in Figure 5.16.

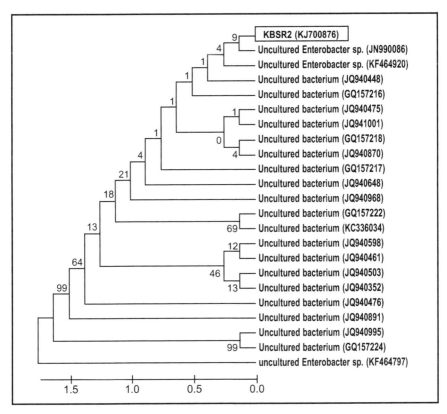

**FIGURE 5.16** Phylogenetic relationship of KBSR2 and other closely related uncultured bacterium based on 16S rDNA sequencing. The tree was generated using the neighbour joining percentage values are given at the nodes.

## 5.18.2   PHYLOGENY OF KBSR3

The homologous search result of KBSR3 demonstrated 98% similarity of 16S rDNA sequence with the unidentified marine bacterioplankton clone 16S ribosomal RNA gene, partial sequence. The phylogenetic tree was constructed from the sequence data by the Neighbour-Joining method which showed the uncultured bacterium (GenBank KC001435) to have 98% homology with KBSR3 and represented its closest phylogenetic neighbour. The phylogenetic tree of KBSR3 is shown in Figure 5.17.

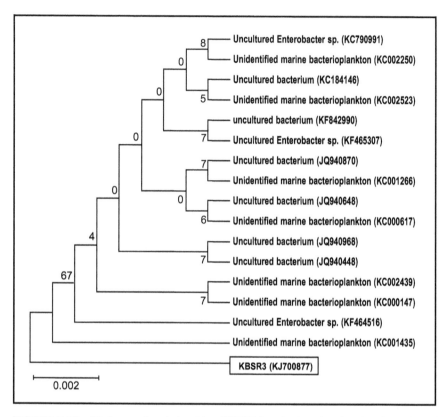

**FIGURE 5.17** Phylogenetic relationship of KBSR3 and other closely related uncultured bacterium based on 16S rDNA sequencing. The tree was generated using the neighbour joining method. The data set was re-sampled 1000 times by using the bootstrap option and percentage values are given at the nodes.

## REFERENCES

Birnboim, H. C.; Doly, J. A Rapid Alkaline Extraction Procedure for Screening Recombinant Plasmid DNA. Nucleic Acids Res. **1979,** *7,* 1513–1523.

Bollet, C.; et al. A Simple Method for the Isolation of Chromosomal DNA from Gram Positive or Acid-fast Bacteria. *Nucleic Acids. Res.* **1991,** *19,* 1995.

Brady, S. F.; Chao, C. J.; Clardy, J. Long-chain N-acyltyrosine Synthases from Environmental DNA. *Appl. Environ. Microbiol.* **2004,** *70,* 46865–46870.

Castro, A. P. de; et al. Construction and Validation of Two Metagenomic DNA Libraries from Cerrado Soil with High Clay Content. *Biotechnol. Lett.* **2011,** *33,* 2169–2175.

Cieslinski, H.; et al. Identification and Molecular Modeling of a Novel Lipase from an Antarctic Soil Metagenomic Library. *Pol. J. Microbiol.* **2009,** *58* (3), 199–204.

Daniel, R. The Metagenomics of Soil. *Nature Microbiol.* **2005**, *3*, 470–478.

Dijkmans, R.; et al. Rapid Method for Purification of Soil DNA for Hybridization and PCR Analysis. *Microb. Releases.* **1993**, *2*, 29–34.

Frostegard, A.; et al. Quantification of Bias Related to the Extraction of DNA Directly from Soils. *Appl Environ Microbiol.* **1999**, *65*, 5409–5420.

Gray, J. P.; Herwig, R, P. Phylogenetic Analysis of the Bacterial Communities in Marine Sediments. *Appl Environ Microbiol.* **1996**, *62* (11), 4049–4059.

Harry, M.; et al. Evaluation of Purification Procedures for DNA Extracted from Organic Rich Samples: Interference with Humic Substances. *Analusis* **1999**, *27*, 439–442.

Henne, A.; et al. Screening of Environmental DNA Libraries for the Presence of Genes Conferring Lipolytic Activity on *Escherichia coli. Appl. Environ. Microbiol.* **2000**, *66*, 3113–3116.

Hilger, A. B.; Myrold, D. D. Method for Extraction of Frankia DNA from Soil. *Agric. Ecosys. Environ.* **1991**, *34*, 107–113.

Hong, K. S.; et al. Selection and Characterization of Forest Soil Metagenome Genes Encoding Lipolytic Enzymes. *J. Microbiol. Biotechnol.* **2007**, *17* (10), 1655–1660.

Jackson, C. R.; et al. A Simple, Efficient Method for the Separation of Humic Substances and DNA from Environmental Samples. *Appl. Environ. Microbiol.* **1997**, *63*, 4993–4995.

Jiang, X.; et al. Identification and Characterization of Novel Esterases from a Deep-sea Sediment Metagenome. *Arc. Miocrobiol.* **2012**, *194*, 207–214.

Jimenez, D. J.; et al. A Novel Cold Active Esterase Derived from Colombian High Andean Forest Soil Metagenome. *World J. Microbiol. Biotechnol.* **2012**, *28*, 361–370.

Kennedy, J.; Marchesi, J. R.; Dobson, A. D. W. Marine Metagenomics: Strategies for the Discovery of Novel Enzymes with Biotechnological Applications from Marine Environments. *Micro. Cell.Factor.* **2008**, *7*, 27.

Lee M. H.; et al. Isolation and Characterization of a Novel Lipase from a Metagenomic Library of Tidal Flat Sediments: Evidence for a New Family of Bacterial Lipases. *Appl Environ Microbiol.* **2006b**, *72*, 7406–7409.

Manjula, A.; et al. Comparison of Seven Methods of DNA Extraction from Termitarium for Functional Metagenomic DNA Library Construction. *J. Scientific. Industrial Research.* **2011**, *70*, 945–951.

Miller, D. N.; et al. Evaluation and Optimization of DNA Extraction for Soil and Sediment Samples. *Appl. Environ. Microbiol.* **1999**, *65*, 4715–4724.

Porteous L. A.; et al. An Improved Method for Purifying DNA from Soil for Polymerase Chain Reaction Amplification and Molecular Ecology Applications. *Mol. Ecol.* **1997**, *6*, 787–791.

Purdy, K. J.; et al. Rapid Extraction of DNA and rRNA from Sediments by a Novel Hydroxyapatite Spin-column Method. *Appl. Environ. Microbiol.* **1996**, *62*, 3905–3907.

Rajendhran, J.; Gunasekaran, P. Strategies for Accessing Soil Metagenome for Desired Applications. *Biotechnol. adv.* **2008**, *26*, 576–590.

Riesenfeld, C. S.; Goodman, R. M.; Handelsman, J. Uncultured Soil Bacteria are a Reservoir of New Antibiotic Resistance Genes. *Environ Microbiol.* **2004**, *6*, 981–989.

Schneegurt, M. A.; Dor, S. Y.; Kulpa, C. F. Jr. Direct Extraction of DNA from Soils for Studies in Microbial Ecology. *Curr Issues Mol Biol.* **2003**, *5*, 1–8.

Selenska, S.; Klingmüller, W. Direct Detection of Nif-gene Sequences of *Enterobacter Agglomerans* in Soil. *FEMS Microbiol. Lett.* **1991**, *80*, 243–246.

Torsvik, V.; et al. Novel Techniques for Analysing Microbial Diversity in Natural and Perturbed Environments. *J. Biotech.* **1998,** *64,* 53–62.

Tsai, Y.; Olson, B. H. Rapid Method for Direct Extraction of mRNA from Seeded Soils. *Appl. Environ. Microbiol.* **1991a**, *57,* 765–768.

Zhang, T.; Han, W. J.; Liu, Z. P. Gene Cloning and Characterization of a Novel Esterase from Activated Sludge Metagenome. *Microbial. Cell. Factories* **2009,** *8,* 67.

Zhou, J.; Bruns, M. A.; Tiedje, J. M. DNA Recovery from Soils of Diverse Composition. *Appl. Environ. Microbiol.* **1996,** *62,* 316–322.

# CHAPTER 6

# FUNCTIONAL APPROACH FOR METAGENOMIC LIBRARY CONSTRUCTION

## CONTENTS

Metagenomic DNA (mgDNA) library is constructed using Pushpam et al. (2011) method. Soil contains diverse microbial populations of both culturable and non-culturable bacteria, therefore, mgDNA (3.8 µg/g soil) was isolated by the indirect extraction method. A small-insert metagenomic library was constructed in the cloning and expression vector pET-32a.

## 6.1   RESTRICTION DIGESTION OF MGDNA

*Sau*3AI recognizes the sequence GATC and generates fragments with 5′-cohesive termini, which contain the tetranucleotide sequence GATC as the cohesive termini of *Bam*HI. An amount of 5 µg of mgDNA is partially digested with FastDigest *Sau*3AI. Setting of restriction digestion reaction as described below:

| Components | Volume (µl) |
|---|---|
| mgDNA | 10.0 (5 µg) |
| 10x buffer | 3.0 |
| *Sau* 3AI (10 U/µl) | 5.0 |
| Deionized water | 12.0 |
| Total volume | 30.0 |

The reaction mixture is incubated at 37°C for 5 min and then the enzyme is inactivated by heating at 70°C for 10 min. The digested DNA fragments are resolved on 0.7% Agarose gel in 1x TAE. mgDNA 5.0 µg was partially digested with *Sau*3AI. DNA fragments ranging from 0.5 to 3.0 kb were resolved in 0.7% (w/v) agarose gel electrophoresis (Fig. 6.1).

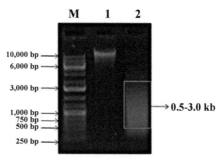

**FIGURE 6.1**   Agarose gel electrophoresis of extracted metagenomic DNA. Lane M–1 kb DNA ladder; lane 1—metagenomic DNA extracted from soil; lane 2—restriction digestion of mgDNA.

## 6.2   TRANSFORMATION OF *E. coli* BL21 COMPETENT CELLS

After the separation by Agarose gel electrophoresis (Fig. 6.1), the gel is examined under UV transilluminator and the gel slice containing DNA fragments ranging about 0.5–3 kb is separated using sterile scalpel. The excised gel is mixed with 3 volume of QG buffer, incubated at 50°C until the gel slice got dissolved completely. For binding of DNA, the sample is applied to QIAquick column, and centrifuged for 1.5 min at 13,000 rpm. The flow-through is discarded and the column is placed back in the same collection tube. QG buffer 500 µl is added to the column and centrifuged for 1.5 min. A volume of 750 µl PE buffer is added to the column for washing and centrifuged for 1.5 min. The column is centrifuged for an additional 1 min at 13,000 rpm to remove the residual buffer. The QIAquick column is placed into a clean 1.5-ml microfuge tube. To elute DNA, 40 µl of deionized water (preheated at 65°C) is added to the center of the QIAquick membrane. The column is allowed to stand for 1 min and then centrifuged for 1.5 min. The recovered DNA fragments are used for library construction.

## 6.2.1 ISOLATION OF PET-32A AND TRANSFORMATION OF E. coli BL21 COMPETENT CELLS

Natural *Escherichia coli* strains often carry plasmids specifying resistance to antibiotics. All plasmid used as cloning vectors contain three common features; an origin of replication, a selectable marker gene, and a multicloning site. pET-32a (Fig. 6.2) is the most powerful system yet developed for the cloning and expression of recombinant proteins in *E. coli*. The target genes are cloned in pET plasmids under the control of strong bacteriophage T7 transcription and translation signals. The expression is induced by providing a source of T7 polymerase in the host cell. T7 RNA polymerase is so selective and active that almost all the cell's resources are converted to target gene expression; the desired product can comprise more than 50% of the total cell protein within few hours after induction. Another important benefit of the system is its ability to maintain target genes transcriptionally silent in the uninduced state.

Target genes are initially cloned using hosts that do not contain the T7 RNA polymerase gene, thus eliminating plasmid instability due to the production of proteins being potential toxic to the host cell. Once established in a non-expression host, plasmids are transferred into expression host containing a chromosomal copy of the T7 RNA polymerase gene under the lacUV5 control, and expression is induced by the addition of IPTG.

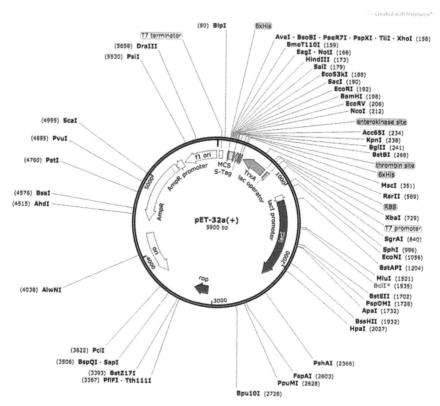

**FIGURE 6.2** pET-32a expression vector system.

The plasmid pET-32a is isolated from the overnight grown culture of *E. coli* by alkali lysis method (Birnboim and Doly, 1979) with minor modifications (Table 6.1). The isolated pET-32a plasmid DNA was electrophoresed along with 1.0-kb DNA ladder. Size of the isolated plasmid DNA was found to be 5.9 kb. The gel is presented in Figure 6.3. The purity of DNA was calculated by taking optical density (OD) at 260 and 280 nm in a UV/VIS spectrophotometer (Thermoscientific, USA). The purity of isolated plasmid DNA samples was 1.9. The DNA yield was calculated using the OD at 260 nm.

**TABLE 6.1** Yield and Purity of Isolated pET-32a Plasmid DNA.

| Sample | DNA yield (μg/μl) | Optical density at A260/A280 |
|---|---|---|
| pET-32a | 0.14 | 0.14 |

## 6.2.2   RESTRICTION DIGESTION OF PET-32A

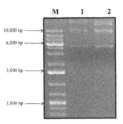

**FIGURE 6.3**   pET-32a plasmid DNA isolation. Lane M—1 kb DNA ladder; lane 1 to 2—extracted plasmid DNA.

*Bam*HI is a type II restriction endonuclease enzyme, which recognizes a short sequence of 6 bp of DNA. *Bam*HI binds at the recognition sequence 5′-GGATCC-3′ and cleaves these sequences just after the 5′-guanine on each strand. This cleavage results the sticky ends that are 4 bp long. pET-32a plasmid DNA 5 µg is digested with *Bam*HI. The setting of the digestion reaction is presented below:

| Components | Volume (µl) |
|---|---|
| pET-32a plasmid | 20.0 (5 µg) |
| 10x buffer | 3.0 |
| *Bam*HI (10U/µl) | 1.0 |
| Deionized water | 6.0 |
| Total volume | 30.0 |

The reaction mixture is incubated at 37°C for 2 h and the enzyme inactivated by heating at 70°C for 10 min. The linearized plasmid DNA is dephosphorylated by adding 1 unit of alkaline phosphatase (CIAP, Fermentas, USA), and incubated at 37°C for 1 h to prevent self-ligation. The digested pET-32a plasmid DNAs are resolved on 1% Agarose gel in 1x TAE. The restricted and dephosphorylated plasmid DNAs are purified using QIAquick Gel Extraction kit.

pET-32a was digested using *Bam*HI restriction enzyme. After restriction digestion of pET-32a single DNA band sized ~5.9 kb (Fig. 6.4) was observed in 1% agarose gel electrophoresis.

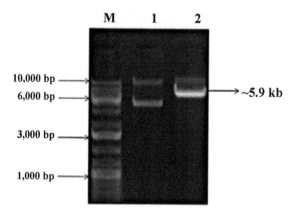

**FIGURE 6.4**   Restriction digestion of the plasmid pET-32a DNA. Lane M—1 kb DNA ladder; lane 1—undigested plasmid DNA; lane 2—digested plasmid DNA.

## 6.2.3   LIGATION

Purified mgDNA fragments ranging about 0.5–3.0 kb are mixed with the digested and dephosphorylated pET-32a plasmid vector DNA in 3:1 ratio. The ligation reaction is set as described below:

| Components | Volume (10 µl) |
|---|---|
| Ligation buffer | 5.0 |
| pET-32a vector (50 ng) | 1.0 |
| mgDNA fragments | 3.0 |

The reaction mixture was incubated overnight at 16°C.

## 6.2.4   TRANSFORMATION OF E. COLI BL21 (DE3) COMPETENT CELLS

The competent cell BL21 (DE3) pLysS is a chemically competent *E. coli* BL21 (DE3) pLysS cell. The BL21(DE3) pLysS strain contains the T7 RNA polymerase gene controlled by the lac UV5 promoter in its chromosomal DNA and the T7 lysozyme gene in the pLysS plasmid. T7 RNA polymerase is expressed upon addition of isopropyl-1-thio-β-D-galactoside (IPTG), which induces a high-level protein expression from T7 promoter driven expression vector (pET). *E. coli* BL21(DE3) pLysS strain is a derivative of *E. coli* strain and lacks both the ion protease and the ompT membrane protease, which may degrade expressed proteins. The *E. coli* BL21(DE3) competent cells are prepared by following Sambrook and Russel (2001) method with some modifications.

## 6.3 FUNCTIONAL SCREENING OF METAGENOMIC LIBRARIES FOR THE LIPASE GENE

The probability (hit rate) of identifying a certain gene depends on the various factors that are inextricably linked to each other like the host–vector system, size of the target gene, its abundance in the source metagenome, the assay method, and the efficiency of heterologous gene expression in a surrogate host. Functional screening of metagenomic libraries is a powerful approach to identify and assign function of novel genes and their encoded proteins without any prior sequence knowledge. The recombinants are screened for lipolytic activity on tributyrin agar plates supplemented with 1% (v/v) tributyrin and incubated at 37°C for 48–72 h. Lipolytic clones are selected based on the zone of clearance around the colony.

All 87,000 clones were screened for lipolytic activity on tributyrin agar plate medium supplemented with 1% tributyrin + ampicillin (100 μg/ml) + IPTG (40 μg/ml). The plates were incubated for 48–72 h. After screening the 87,000 recombinant clones for lipolytic activity, one clone was found to exhibit zone of clearance around the colony after 48 h of incubation at 37°C and the same was designated as KBSplip1. The recombinant clone showing the lipolytic activity is presented in Figure 6.5.

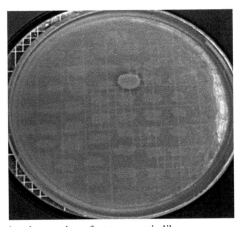

**FIGURE 6.5**  Functional screening of metagenomic library.

The transformation of recombinant and non-recombinant plasmid is done by following Sambrook and Russel (2001) method. Metagenomic DNA fragments of 0.5–3.0 kb size were ligated into the restricted pET-32a plasmid vector using T4 DNA ligase enzyme. The ligated mixture was transformed to *E. coli* BL21(DE3) host cells by the heat shock method. The transformation reaction mixture was plated on LBA + ampicillin (100 µg/ml) + 6 µl of 200 mM IPTG plates and incubated overnight at 37°C. The library consisted of approximately 87,000 clones.

## 6.4 ISOLATION AND RESTRICTION DIGESTION OF RECOMBINANT PLASMID

A loopful of the above lipase producing recombinant bacterial culture is inoculated into 50 ml Luria broth (LB) medium supplemented with ampicillin (100 µg/ml) and incubated overnight at 37°C. The recombinant plasmid is isolated from the lipase producing positive clones by alkali lysis method followed by Birnboim and Doly (1979) with minor modifications as mentioned earlier (Table 6.2). The restriction digestion of recombinant plasmid is done using *Bam*HI restriction enzyme. The plasmid KBS-plip1 DNA (5 µg) is digested with *Bam*HI. The setting of the digestion reaction is presented below:

| Components | Volume (µl) |
|---|---|
| KBS-plip1 plasmid | 20.0 (5 µg) |
| 10x buffer | 3.0 |
| *Bam*HI (10 U/µl) | 1.0 |
| Deionized water | 6.0 |
| Total volume | 30.0 |

**TABLE 6.2**  Yield and Purity of Isolated Recombinant Plasmid DNA (KBS-plip1).

| Sample | A260 | A280 | A260/A280 | DNA yield (µg/ml) |
|---|---|---|---|---|
| 1 | 0.317 | 0.162 | 1.95 | 79.25 |
| 2 | 0.301 | 0.157 | 1.91 | 75.25 |

The recombinant clone showing the lipolytic activity was grown overnight in LB medium containing ampicillin (100 µg/ml) at 37°C. The isolated plasmid DNA was electrophoresed along with 1.0 kb DNA ladder.

The size of the recombinant plasmid was about 0.9 kb greater than the non-recombinant plasmid. The yield and purity of the isolated plasmid DNA was found to be 84.0 µg/ml and 1.91, respectively.

The recombinant KBS-plip1 digested with *Bam*HI restriction enzyme showed two DNA bands of 5900 bp and ~900 bp for pET-32a plasmid vector and insert, respectively. The restriction digestion of recombinant KBS-plip1 plasmid is shown in Figure 6.6(a) and (b).

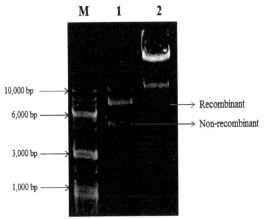

**FIGURE 6.6 (a)**    Isolation of recombinant plasmid. Lane M–1 kb DNA ladder; lane 1—nonrecombinant pET-32a; lane 2—recombinant pET-32a.

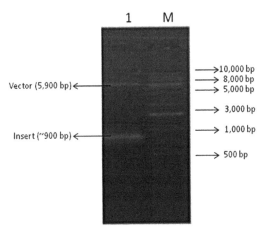

**FIGURE 6.6 (b)**    Restriction digestion of recombinant plasmid. Lane M–1 kb DNA ladder; lane 1—recombinant KBS-plip1 plasmid digested with *Bam*HI restriction enzyme.

## 6.5  SEQUENCING OF CLONED DNA

The nucleotide sequence of the inserted DNA from the positive clone is sequenced using ABI PRISM-3100 sequencer (Applied Biosystems, Foster city, CA, USA) with the T7 promoter primer. Sequencing reaction is performed by using BigDye® Terminator v3.1 Cycle Sequencing Kit (Applied Biosystem). In a 0.2 ml PCR tube, 5.0 μl milli Q water, 1.5 μl 5x dilution buffer (supplied in the kit), 0.5 μl (10 pmol/μl) of T7 promoter sequencing primer, 1.0 μl 2.5x Ready reaction mixture (containing DNA polymerase, dNTPs, and dye-labeled ddNTPs), and 2.0 μl target DNA to be sequenced (50 ng/μl of plasmid DNA) are added. Primers used in DNA sequencing is given below:

### 6.5.1  PRIMER NAME PRIMER SEQUENCE (5′-3′)

T7 Promoter TAATACGACTCACTATAGGG
 PCR conditions used for the sequencing are given below:

| Steps | Conditions |
|---|---|
| Initial denaturation | 96°C for 1 min |
| Denaturation | 96°C for 10 s |
| Hybridization | 50°C for 5 s |
| Elongation | 60°C for 4 min |
| Cycles | 25 |

After PCR, the tubes are stored at 4°C till further purification. To remove the un-incorporated dye terminators and the primer, 12 μl master mix-I (2.0 μl 125 mM EDTA, 10 μl water), followed by 52 μl of master mix-II (2.0 μl–3M NaOAc; pH 4.6, 50 μl 95% ethanol) are added to the sequencing product. The contents are mixed by gentle tapping, incubated at room temperature for 15 min, and centrifuged at 10,000 rpm for 15 min. The supernatant is discarded, the pellet washed twice with 100 μl of 70% ethanol. The pellet is air-dried for 5 min and then dissolved in 24 μl Hi-Diformamide (Applied Biosystem, USA). The purified product (12 μl) is loaded into the 96-well plate of the auto-sampler in the sequencer. The sequence data is aligned and compared with the published sequences obtained from the GenBank database using Seq Scape v 5.2. The recombinant plasmid KBS-plip1was sequenced using T7 promoter primer. The DNA sequence of the cloned gene in pET-32a was about 0.9 kb. However, the clone KBS-plip1 was found to contain an open reading frame (ORF) of 891 bp that was capable of encoding a protein with a predicted molecular mass of 40 kDa. The insert sequence retrieved from the recombinant showed 99% similarity with the lipase producing bacterium KC182797. The nucleotide sequence of lipase coding gene cloned in KBS-plip1.

## 6.5.2 GENBANK SUBMISSION

The nucleotide sequence of KBS-plip1 is deposited in the GenBank database. The cloned lipase gene KBS-plip1 sequence from the soil metagenome was deposited in the GenBank under accession number KF743145.

## 6.6 SEQUENCE ANALYSIS AND PHYLOGENETIC TREE CONSTRUCTION

A complete ORF consisting of 891 nucleotides is identified by an ORF finder. The BLASTP program is used to screen the amino acid sequence database. Multiple sequence alignments are constructed using the CLUSTAL-W program and visually examined with BoxShade Server program. To determine the evolutionary relationship of the environmental esterase with established bacterial lipolytic enzymes, KBS-plip1 sequence is compared to the reference bacterial lipolytic enzyme from protein sequence database (NCBI), using neighbour joining phylogenetic analysis. One thousand bootstrap replications are performed using the program MEGA 4.0.

## 6.7 ANALYSIS OF THE CLONED LIPASE GENE

To analyse the cloned lipase gene, the nucleotide sequence of KBS-plip1 is translated into amino acid sequence using ExPASy translate tool. Multiple sequence alignment for lipase is performed with the other known lipase sequences in the NCBI database using BLAST P algorithm.

## 6.8  PHYLOGENY OF RECOMBINANT PROTEIN

The phylogenetic tree was generated using the neighbour joining method. The data set was re-sampled 1000 times by using the bootstrap option and percentage values are presented at the nodes.

## 6.9  AMINO ACID SEQUENCE OF RECOMBINANT PROTEIN

The nucleotide sequence of KBS-plip1 was translated into the amino acid sequence using (http://expasy.org/translate) translate tool. The protein sequence for KBSplip1 was found to be 291 amino acid residues (Fig.6.7).

```
MSQQQLQSIIQMLKSQPIAGKPSIAETRAGFEQMAAMFPVEADVKSEPVNAGGVKSEWVTAPGADA
GRAVLYLHGGGYVIGSISTHRASAGRISRAAKARVLVIDYRLAPEHPFPAAVEDSVAAYRWMLSTG
LKPSRIAVAGDAAGGGLTVATLVAIRDAKLPVPAAGVALSPWVDMEGVGDSMKTKAAVDPMVSKDG
LIEMAKAYLGGNDTRTPLAAPLYADLAGLPPLLIQVGTAETLLDDSTRLAERARKAGVKVTLEPWE
NMVHVFQIFASILDEGQQAIDKIGAFIRANAE
```
**FIGURE 6.7**   Protein sequence of cloned lipase gene.

## 6.10  MULTIPLE SEQUENCE ALIGNMENT

Analysis of the insert DNA sequence as described above, revealed an ORF of 891 bp with ATG as the start codon and TAA as the termination codon. The deduced amino acid sequence of the lipase/esterase comprises 296 amino acids residues and an estimated molecular mass of 32,560 kDa. Multiple sequence alignment of this lipase was performed with other known lipase gene sequences in the NCBI database. The amino acid sequence of this KBS-plip1 lipase gene displayed 23–55% sequence similarity with other lipase genes present in NCBI database. KBS-plip1 showed 55% amino acid sequence homology with the uncultured bacterium lipase (accession 3V9A) and hormone-sensitive lipase (HSL) derived from the metagenomic library (accession 3FAK), 43% with HSL from metagenomic library (accession 3DNM), 33% with carboxyl-esterase from the uncultured bacterium (accession 2C7B), and 33% with esterase from the bacterium accession 1LZL. KBS-plip1 lipase gene was classified based on the work from Jaeger et al. (1999), which compared the amino acid sequence with that of HSL family. The red boxes in multiple sequence alignment showed the highly conserved region in the HSL family. The HSL family conserved HGGG motif (amino acids 23–26) was found upstream of the active-site conserved motif in KBS-plip1. HSL contains the lipase-conserved catalytic consensus pentapeptide GDSAG. The catalytic triad DPM (disintegrations per minute, amino acids 145–147) and HVF (human ventricular fibroblasts, amino acids 228–230) are also the conserved sequences of HSL family and present in KBS-plip1 amino acid sequence. The results of multiple sequence alignment showed KBS-plip1 belonging to the HSL family (Fig. 6.8).

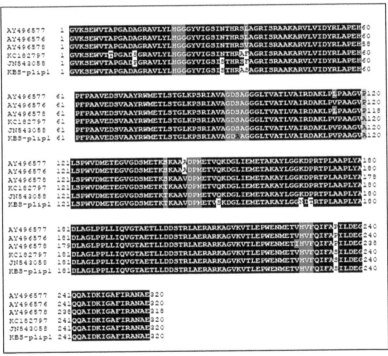

**FIGURE 6.8** Multiple sequence alignment of cloned lipase gene from soil metagenome.

## 6.11 STORAGE OF RECOMBINANT PLASMID

*Escherichia coli* BL21(DE3) has full functional endonuclease and recombinase pathways. As a result, whatever plasmids are stored in *E. coli* BL21(DE3), get damaged or rearranged after some time. Therefore, the isolated recombinant plasmids are transferred into *E. coli* DH5α and stored at −70°C. The isolated and purified recombinant plasmid DNA is also stored at −20°C.

## REFERENCES

Birnboim, H. C.; Doly, J. A Rapid Alkaline Extraction Procedure for Screening Recombinant Plasmid DNA. Nucleic Acids Res. **1979,** 7, 1513–1523.

Jaeger, K. E.; Dijkstra B. W.; Reetz, M. T. Bacterial Biocatalysts: Molecular Biology, Three Dimensional Structures, and Biotechnological Applications of Lipases. *Annu. Rev. Microbiol.* **1999,** *53,* 315–351.

Pushpam, P. L.; Rajesh. T.; Gunasekaran, P. Identification and Characterization of Alkaline Serine Protease from Goat Skin Surface Metagenome. *AMB Express.* **2011,** *1,* 1–10.

Sambrook. J.; Russel, D. W. *Molecular Cloning: A Laboratory Manual,* 3rd ed.; Cold Spring Harbor Laboratory Press: New York, 2001.

# CHAPTER 7

# OVEREXPRESSION OF RECOMBINANT PROTEIN

## CONTENTS

A single colony of *Escherichia coli* BL21(DE3) cells harbouring expression constructs is inoculated in 5 ml Luria broth (LB) medium containing 100 μg/ml ampicillin and grown overnight at 37°C with constant shaking. Out of the culture, 1% (v/v) is inoculated in 500 ml of LB medium supplemented with 100 μg/ml of ampicillin. The culture is grown at 37°C until an absorbance of 0.4–0.6 at OD600 is achieved. The culture is then induced with 0.1–0.5 mM of isopropyl-β-D-thiogalactopyranoside (IPTG) for 5 h with incubation at 30°C. The cells are harvested by centrifugation at 10,000 rpm for 10 min. The cell pellet is stored and applied to metal chelating Ni-NTA affinity chromatography (Bio-Rad) for protein purification.

## 7.1   PURIFICATION OF EXPRESSED PROTEIN

The solubilized protein was purified on Ni-NTA Affinity Chromatography. The recombinant protein contained 6xHis-tag to allow highly homogeneous purification by a Ni-NTA column.

The cell lysate was loaded in Ni-NTA column. In the initial purification, the 6xHis-tagged protein was eluted with the background proteins of 0–20 mM imidazole buffer. After washing off the background proteins with 50 mM of imidazole buffer, a homogeneous protein was eluted with 200 mM imidazole. Eluted and purified protein was analysed by electrophoresis in 12% sodium dodecyle sulphate (SDS) polyacrylamide gel.

Optimization of protein purification is based on several factors, including the amount of 6xHis-tagged protein required and the expression level (Hochuli et al., 1988). The imidazole rings in the histidine residues of the 6xHis-tag bind to the nickel ions immobilized by the NTA groups on the matrix. Imidazole itself can also bind to the nickel ions and disrupt the binding of dispersed histidine residues in non-tagged background proteins. At low imidazole concentration, non-specific-specific, low affinity binding of background proteins is prevented, while 6xHis-tagged proteins still bind strongly to Ni-NTA matrix. Therefore, adding imidazole to the lysis buffer leads to greater purity in fewer steps. Since the interaction between Ni-NTA and 6xHis-tag of the recombinant protein does not depend on tertiary structure, proteins can be purified either under native or denaturing conditions.

### 7.1.1   PROTOCOL

The cell pellet is resuspended in 1 ml lysis buffer (pH 7.4), vortexed for homogeneity. The cell suspension is disrupted using an ultrasonic homogenizer (Sonic Ruptor, OMNI International, India) five times for 30 s with 20-s interval and centrifuged at 13,000 rpm for 30 min. The supernatant is transferred on Ni-NTA columns with 100 μl agarose beads. Before adding the supernatant, the agarose beads are washed with two column volume of the wash buffer. The Ni-NTA column is kept at 4°C for 2 h after adding the supernatant to the column to allow the protein to bind to the beads. After binding the column is washed ten times with wash buffer. The elution buffer 200 μl is added to the column followed by incubation for 30 min to collect the bound protein. Flow through, wash, and elution are collected. The samples are loaded on 12% SDS-PAGE.

The recombinant pET-32a was expressed in *E. coli* BL21(DE3) host cells. Same were inoculated in LB medium containing ampicillin 100 μg/ml at 37°C until an absorbance level of 0.4–0.6 at OD600 is reached. The culture was induced with 0.5 mM IPTG for 5 h at 30°C. The expression of

KBS-plip1 in *E. coli* BL21 (DE3) was achieved without extensive optimization of cultivation and induction conditions, which indicates that KBS-plip1 was inherently amenable to over-expression in *E. coli*. The expressed protein bands were analysed in 12% SDS polyacrylamide gel electrophoresis. Cells bearing KBS-plip1 recombinant plasmid showed expression of product with in size of approximately 40 kDa, while there was no expression of protein in the case where non-recombinants and recombinants were grown without IPTG. Significant portion of the expressed protein was presented in both pellet as well as supernatant.

The solubilized protein was purified on Ni-NTA Affinity Chromatography. The recombinant protein contained 6xHis-tag to allow highly homogeneous purification by a Ni-NTA column.

The cell lysate was loaded in Ni-NTA column. In the initial purification, the 6xHis-tagged protein was eluted with the background proteins of 0–20 mM imidazole buffer. After washing off the background proteins with 50 mM of imidazole buffer, a homogeneous protein was eluted with 200 mM imidazole. Eluted and purified protein was analysed by electrophoresis in 12% SDS polyacrylamide gel.

## 7.1.2  SDS POLYACRYLAMIDE GEL ELECTROPHORESIS

Reagents
  12% acrylamide-bisacrylamide resolving gel
  5% acrylamide-bisacrylamide resolving gel
  1x Tris Glycine buffer (pH 8.3)
  1x gel loading buffer

12% acrylamide-bisacrylamide resolving gel:

| Components | Volume (ml) |
| --- | --- |
| Distilled water | 1.6 |
| 30% acrylamide | 2.0 |
| 1.5 M Tris (pH 8.8) | 1.3 |
| 10% SDS | 0.05 |
| 10% Ammonium per sulphate | 0.05 |
| TEMED | 0.002 |
| Total volume | 5.0 |

5% acrylamide-bisacrylamide stacking gel:

| Components | Volume (ml) |
| --- | --- |
| Distilled water | 1.40 |
| 30% acrylamide | 0.33 |
| 1.5 M Tris (pH 8.8) | 0.25 |
| 10% SDS | 0.02 |
| 10% Ammonium per sulphate | 0.02 |
| TEMED | 0.002 |
| Total volume | 2.0 |

5x Tris Glycine buffer (pH 8.3)—Stock buffer 25 Mm Tris:
   250 mM glycine
   0.1% SDS
   1x Tris Glycine buffer (pH 8.3) was used as working running buffer
   1x SDS gel loading buffer
   50 mM Tris.Cl (pH 6.8)
   100 mM dithiothreitol
   2% SDS
   0.1% bromophenol blue
   10% glycerol

   SDS polyacrylamide gel electrophoresis is carried out with or without reduction of proteins by β-mercaptoethanol as described by Laemmli345. Briefly 50 µg of crude proteins or purified lipase is loaded into the well of 12% separating gel containing 5% glycerol. Electrophoresis is carried out at a constant current of 15 mA until the dye front (bromophenol blue) reached the bottom of the gel. Protein bands are visualized by staining with 0.1% Commassie Brilliant Blue R-250 in methanol: acetic acid: water (4:1:5 v/v/v). The destained gels are scanned in a Biospectrum 500 Imaging system, India. Mobility of the purified protein is compared with the protein molecular weight marker; 10–25 kDa, Thermo Scientific, USA. Molecular weight of the unknown protein is calculated using the RF value of the protein and the dye front.

## 7.2   DETERMINATION OF ACTIVE LIPASE USING ZYMOGRAPHIC STUDY

Lipase zymogram is carried out as described by Glogauer et al. (2011). Briefly, tributyrin (1.0% v/v) is dissolved in a 1.5 M Tris-HCl buffer (pH

8.8) and copolymerized with 12% (w/v) acrylamide, 0.32% (w/v) bis-acrylamide in order to make a running gel. Then 4% (w/v) acrylamide, 0.11% (w/v) bisacrylamide, and 0.5 M Tris-HCl (pH 6.8) without the substrate is used for the stacking gel, and then poured into a mini gel caster (BioRad). The protein sample (15 μg) is prepared by a zymogram sample buffer (0.5 M Tris-HCl, pH 6.8, 20% glycerol, and 0.5% bromophenol blue). Samples are prepared and loaded into the wells and electrophoresed in the cold room at the constant current of 12 mA. After the electrophoresis, the gel is incubated for 20 min at 4°C on a rotary shaker in 50 mM sodium phosphate buffer (pH 8.0). The gel is incubated in zymogram reaction in the same buffer (pH 8.0) at 37°C for 24 h. The gel is stained with Comassie Brilliant Blue (0.5%) for 8 h and then destained in the fixed volume (100 ml) of the destaining solution, which contained 10% methanol and 5% acetic acid for a limited period. For quantification, the density of the destained bands on the zymograms is analysed by video densitometry using 1D ver. 97.04 (VilberLourmat, France) image analyser.

The expressed and purified protein was analysed in 12% SDS-polyacrylamide gel. The size of the protein was about 40 kDa as shown in Figure 4.36. Zymographic analysis was carried out using tributyrin as the substrate. The band resolved at 40 kDa region reveal that the purified enzyme was active (Fig. 7.1 and 7.2).

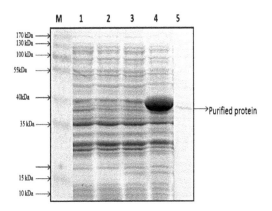

**FIGURE 7.1**   SDS-PAGE of *E. coli* BL-21(DE3) plasmid DNA. Lane M—Page Ruler Pre-stained protein marker (10–170 kDa); lane1—uninduced pET-32a; lane 2—induced pET-32a; lane 3—uninduced KBS-plip1; lane 4—induced KBS-plip1 with 0.5 mM isopropyl-β-D-thiogalactopyranoside (IPTG); lane 5—purified KBS-plip1 using Ni-NTA affinity chromatography.

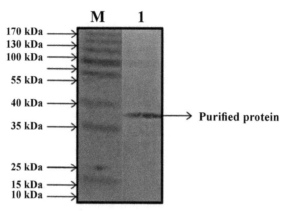

**FIGURE 7.2**  Zymogram of purified KBS-plip1 lipase. Lane M—Page Ruler Pre-stained protein marker (170–10 kDa); lane 1—zymogram of purified KBS-plip1 protein.

## 7.2.1   QUANTIFICATION OF PROTEIN

The protein lipase produced by the recombinant bacterium was estimated as per Lowry's method. The protein concentration present in KBS-plip1 was found to be 0.72 mg/ml.

## 7.3   DETERMINATION OF SPECIFIC ACTIVITY, % YIELD AND PROTEIN FOLD PURIFICATION

Activity of the purified KBS-plip1 recombinant protein was determined using the titrimetric method. The purified protein exhibited a maximum activity 118.6 $U \cdot ml^{-1}$ and specific activity 96.32 $U \cdot mg^{-1}$ (Table 7.1).

**TABLE 7.1**  Specific Activity, % Yield, and Purification Fold of Recombinant Lipase KBS-plip1.

| Samples | Total protein (mg) | Total activity (U) | Specific activity (U/mg) | Yield (%) | Purification fold |
|---|---|---|---|---|---|
| Crude | 600 | 62750 | 43.10 | 100 | 1 |
| Ni-NTA Affinity chromatography | 360 | 34675.2 | 96.32 | 55.2 | 2.23 |

## 7.3.1   DETERMINATION OF LIPASE ACTIVITY OF PURIFIED PROTEIN

Lipase activity is measured in the purified protein using the titrimetric method (Beisson et al., 2000). The activity of the purified KBS-plip1 recombinant protein was determined using the titrimetric method. The purified protein exhibited a maximum activity 118.6 $U \cdot ml^{-1}$ and specific activity 96.32 $U \cdot mg^{-1}$.

## 7.4   HOMOLOGY MODEL AND VALIDATION FOR PROTEIN STRUCTURE PREDICTION

To determine the predicted structure for KBS-plip1, the lipase nucleotide sequence is translated into amino acid sequence using ExPASy Translate tool. The homology modelling is performed using sequence alignment of amino acid sequence against Protein Data Bank (PDB) by BLAST algorithm. The template is selected on the basis of maximum sequence identity with target sequence. Homology model is constructed using Swiss-PDB viewer v. 4.0.1 program. The stereochemical quality is evaluated using Ramachandran plots obtained from the RAMPAGE server. Three-dimensional ribbon model for protein is generated using Molegro-Molecular Viewer v.2.5.0.

To determine the protein structure nucleotide sequence was translated into amino acid sequence using ExPASy tool. The protein sequence contained 296 amino acids and deposited to Bankit with protein id AHF22385. The homology modelling was done using the sequence alignment of the amino acid sequence of KBS-plip1 against the protein sequence data bank. Based on the alignment, a model of KBS-plip1 lipase was made using a bacterial lipase (PDB ID: ABZ9) as template for the model building. Sequence alignment showed both target sequence (KBS-plip1) and template ABZ9 share 72% of sequence identity. The modelled enzyme is a folded dimmer. It consists of 9 β-strands with 7 α-helices. The ribbon model was generated using Molegro-Molecular Viewer v. 2.5.0 (Fig. 7.3).

**FIGURE 7.3**   Ribbon representation of a three-dimensional structure of KBS-plip1 lipase enzyme in Molegro-Molecular Viewer v. 2012 2.5.0 (2D structure).

The enzyme has a catalytic triad, elements of which are borne on loops and best conserved structural features of the fold. The enzyme contains a Ser-centred consensus sequence and a conserved His-Gly dipeptide found in N-terminal domain. The Ramachandran plot (Fig. 7.4) indicated the region of possible angle formations by $\varphi$ and $\psi$ angles. The conventional terms represented the torsion angles on either side of the $\alpha$-carbon in peptides. The Ramachandran plot was divided into three regions: Favoured (95.6%), allowed (4%), and outlier (0.4%). The result is significant as there was high percentage of residues in the favoured region (>95%). This indicates that the built model is of good quality.

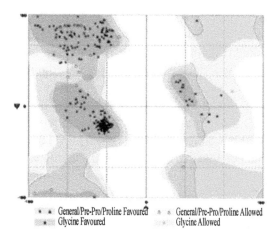

General/Pre-Pro/Proline Favoured    General/Pre-Pro/Proline Allowed
Glycine Favoured                    Glycine Allowed

**FIGURE 7.4**   Ramachandran plot for validation of protein structure prediction; number of residues in the favored region (95.6%); number of residues in the allowed region (4.0%); and number of residues in the outlier region (0.4%).

Recent progress in molecular microbial ecology has revealed that traditional culturing methods fail to represent the scope of microbial diversity in nature. The sequence-based metagenomic approach is a powerful tool to explore the microbial diversity present in soil. Sequencing of the 16S rRNA PCR amplicon is the most common approach for investigating environmental prokaryotic diversity and taxonomic profile of microbial communities based on 16S rDNA. Robe et al. (2003) reported that soil is a complex environment, which appears to be a major reservoir of microbial diversity. Bacteria are the most abundant organisms in soil and can form the largest component of the soil biomass. In this study, the size of 16S rDNA amplicons was found to be approximately 1500 bp in 1% Agarose gel electrophoresis. As per the protocol of Martin-Laurent et al. (2001), quality mgDNA was used to amplify16S rRNA bacterial gene using 27F and 1492R universal primers. Following the protocol of Dugan et al. (2002) the excess dNTPs, primers, polymerase, MgCl2 preset in PCR products were removed by using QIAquick PCR product purification kit and QIAquick gel extraction kit to prevent their inhibitory effect. Unconsumed dNTPs and primers remaining in PCR products mixture, interfere to get positive clones due to insertion of primers into vector. Bacterial 16S rRNA gene library from soil metagenome was constructed using pGEMT-Easy vector. Girija et al. (2013) also used the same system to construct 16S rRNA gene library. T4 DNA ligase with minimal exonuclease activity and nuclease free water were used for ligation reaction. According to Clark (1988) nuclease may degrade the T-overhangs on vector. The ligation was transformed to *E. coli* (DH5α) competent cells. The positive clones were analysed by colony PCR using universal M13F, M13R, and 16S rRNA gene specific 27F, 1492R universal primers as assessed in 1.2% (w/v) agarose gel. The amplicon size was found to be approximately 1.5 kb, but the size was very less in the case of recombinant plasmid. This was due to the presence of the insert DNA in the plasmid. In the case of non-recombinant plasmid, primers amplified only the vector DNA; therefore, the size remained small. The recombinant plasmid DNA from the resultant white colonies of *E. coli* (DH5α) cells was isolated. The purity of the isolated plasmid A260/A280 was between 1.76 and 1.94, suggesting high-quality DNA. The DNA yield ranged from 400 to 2300 ng/μl. The size of recombinant plasmid DNA with the size of 4.5 kb suggested the insertion of 1.5 kb mgDNA fragment into the plasmid (pGEMTEasy vector) having the size of 3 kb. Rondon et al. (2000) reported the similar data to clone soil metagenome. The restriction digestion of recombinant plasmid resulted in two DNA fragments and one in the case of non-recombinant plasmid. This released the insert DNA from the recombinant plasmid vector.

A total of three 16S rRNA gene clones KBSR1, KBSR2, and KBSR3 were sequenced. BLAST search in NCBI database for the sequence homology of 16S rRNA gene exhibited clone KBSR1 to have 97% homology with uncultured bacterium FR687433; clone KBSR2 99% homology with uncultured *Enterobacter* JN990086; and clone KBSR3 98% homology with uncultured bacterioplankton KC001435. In the present investigation, the clones KBSR1, KBSR2, and KBSR3 showed the 16S rDNA sequence homology with the uncultured bacterium supported that 99% of the bacterial population in an environment remain to be uncultured. Schloss and Handelsman reported the same that more than 99% of the microorganisms present in natural environments are not readily culturable. Sagar et al. (2012) stated that bakery wastes usually comprise of a significant proportion of oil remnants, and as such it is quiet arguable to expect a wide range of bacterial strains in these sites, which may utilize oil remnants (lipids) as the sole carbon source. The potential genes were identified using functional screens and had little homology to known genes, which illustrates the enormous potential of soil-based metagenomic libraries. In the present investigation, an attempt was made to construct soil metagenomic library from the samples of the bakery waste dumping site, Tezpur, Assam to identify lipase gene.

According to Daniel (2005) mgDNA constitutes a promising source of novel metabolites. Potential discovery of these compounds requires the construction of good quality metagenomic library and its efficient screening. Torsvik et al. (1998) reported according to reassociation kinetic data, the genetic diversity of the soil metagenome is between 5000 and 5,000,000-fold higher than that of the *E. coli* genome. Such diversity requires improved cloning efficiency so that the clones in the gene library provide an acceptable representation of the entire metagenome.

Urban and Adamczak (2008) reported that culture-independent methods to discover biocatalysts usually require the creation of a library of DNA inserts smaller than 10 kb. It is also required that a relatively large number of clones should be obtained to make up for a small number of clones, which are active on selective substrates. The isolation of large DNA fragments requires mild conditions and consist mainly lysis of microorganism cells. Cloning of large DNA fragments provides a better chance to obtain phylogenetic marker or a gene that encodes the specific function or enzyme. In the present investigation metagenomic DNA was extracted using indirect DNA extraction method to avoid the DNA shearing. mgDNA (3.8 µg/g soil) was used to construct the same. According to Pushpam et al. (2011), the indirect method of DNA isolation yielded higher molecular mass DNA with greater purity then the direct method. The direct DNA extraction method from soil and sediments are hampered by the problems of mechanical shearing due to physical forces imposed on the sample during isolation, such as bead beating. Furthermore, nucleases released during cell lysis may degrade the released DNA188. DNA extraction using direct method contained 61–93% of eukaryotic nucleic acids, which may be due to the partial lysis of indigenous eukaryotic organisms such as fungi, algae, and protozoa, or it may be caused by lysis of residual plant material. However, DNA obtained by indirect method was primarily derived from bacterial cells (>92%) due to the separation from Eukarya by differential centrifugation, which makes it suitable for expression cloning. Direct lysis should be avoided when gene banks are constructed in bacterial hosts. Extracted metagenomic DNA was partially digested with *Sau* 3A enzyme to get the larger DNA fragments.

Various researchers have identified gene(s) EstAS, pELP45, pELP11B, pELP141, pELP81, pELP102, pELP182, EstC23, EM3L1, EM3L2, EM3L3, EM3L4, EM3L6, EM3L7, LipG, Est1, Est2, Est3, Est4, and Est5 from soil metagenomic libraries for the production of lipases/esterases. In the present study, pET-32a was used to construct mgDNA library. The purity of isolated pET-32a plasmid DNA was 1.9 and the yield was calculated to be 0.14 µg/µl. According to Pushpam et al. (2011), pET expression vector system is one of the most widely used systems for the cloning and in vivo expression of recombinant proteins in *E. coli*. The restriction digestion of pET-32a DNA by the endonuclease enzyme yielded fragments ~5.9 kb in Agarose gel electrophoresis. DNA fragments digested with *Bam*HI are compatible with the fragments produced by *Sau*3AI. The partially digested DNA fragments of 0.5–3.0 kb were cloned into the *Bam*HI digested pET-32a plasmid. The difference between the digestion by *Bam*HI and *Sau*3AI is the length in the base-pair recognition site and, consequently, in the frequency of cutting the DNA. *Bam*HI has a 6bp-recognition sequence whereas *Sau*3AI 4bp. Therefore, restriction digestion of mgDNA using *Sau*3AI produced smaller pieces of DNA than *Bam*HI, which would cut more often. Both enzymes produce identical cohesive ends and therefore

DNA pieces of vector and mgDNA were ligated. Metagenomic library was maintained in *E. coli* BL21(DE3) competent cells. Multiple ligation and transformation were done to increase the number of clones in the metagenomic library. The construction of mgDNA library using multiple ligation and transformation was reported to increase the number of clones in the library. The transformation efficiency of the competent cells was $7.6 \times 10^7$ cfu/µg DNA. This transformation efficiency was sufficient to get more number of clones. The functional screening of metagenomic library for the lipolytic activity in tributyrin (1.0%) agar plate supplemented ampicillin (100 µg/ml) and IPTG (40 µg/ml). Out of 87,000 recombinants, one clone exhibited a zone of clearance around the colony after 48 h of incubation at 37°C and the same was considered to a putative recombinant plasmid and designated as KBS-plip1. Jiang et al. (2012) identified and characterized the novel esterase from a deep-sea sediment metagenome using the same.

The recombinant KBS-plip1 showing lipolytic activity was used for plasmid isolation. The recombinant plasmid along with the non-recombinant one was electrophoresed along with 1.0 kb DNA ladder. The size of recombinant plasmid was ~0.9 kb larger than the non-recombinant plasmid. The insert sequence of the lipolytic positive clone KBS-plip1 was determined by DNA sequencing. The recombinant plasmid KBS-plip1 was sequenced using T7 promoter primer. The DNA sequence of cloned gene in pET-32a was 1.0 kb. The ORF was identified via Open Reading Frame Finder. However, the clone KBS-plip1 revealed the presence of an open reading frame of 891 bp. The insert sequence KF743145 retrieved from the recombinant showed significant 99% similarity with lipase producing uncultured bacterium KC182797. The nucleotide sequence of KBS-plip1 lipase was translated into amino acid sequence using the translate tool and consists of 296 amino acid residues. As per Jaeger et al., 1999, on the basis of multiple sequence alignment, KBS-plip1 belongs to family IV (HSL family). According to them lipases/esterases are classified into eight families and KBS-plip1 contains the lipase conserved catalytic consensus pentapeptide GDSAG, Catalytic triad DPM (amino acids 145–147), and HVF (amino acids 228–230). A Ser-centred consensus sequence and a conserved His-Gly dipeptide were found in N-terminal domain of the enzyme. The present investigation suggested that KBS-plip1 belong to hormone-sensitive lipase (HSL) family. According to Langin et al. (1993), the lipolytic enzymes which belong to family IV are significantly similar to mammalian hormone-sensitive lipases (HSL), and most of the enzymes in this family are from psychrophilic and thermophilic bacteria. The mgDNA library was maintained in *E. coli* (DH5α). The library maintained in the expression vector containing host cells of *E. coli* BL21(DE3) shows instability. *E. coli* BL21(DE3) has functional endonuclease and recombinase pathways. As a result, whatever plasmids stored in *E. coli* BL21(DE3) get damaged or rearranged after some time. Therefore, isolated recombinant plasmids were transferred into *E. coli* (DH5α) and stored at −70°C and −20°C.

The recombinant protein in KBS-plip1 overexpressed in the presence of 0.5 mM IPTG for 5 h at 30°C. The calculated molecular weight of KBS-plip1 was 32,560 kDa. However, the size of expressed protein was found to be approximately 40 kDa due to presence of His-tag with recombinant protein with the combination of 0.72 mg/ml. Poly-histidine tag (His-tag) is the most commonly used tag for collecting large amounts of highly purified protein. The hydrophilic and flexible nature of His-tags can often increase the solubility of target proteins and rarely interfere with protein's function. It comprises 6–14 His and is typically fused to the N- or C-terminal end of a target protein. $Ni^{2+}$ shows the highest affinity and selectivity for His-tags and is therefore preferred. The activity of KBS-plip1 expressed and purified esterase was analysed using SDS polyacrylamide gel zymography. The clear band in SDS polyacrylamide gel was obtained in the 40 kDa region reveal that the purified enzyme was active. Glogauer et al. (2011) 346 reported the zymography for novel lipase from soil metagenome showing the clear band in SDS-PAGE. The maximum lipase activity of the purified KBS-plip1 recombinant protein was 118.6 $U \cdot ml^{-1}$ with specific activity of 96.32 $U \cdot mg^{-1}$. The lipase activity of purified KBS-plip1 was determined using titrimetric method.

The structure of KBS-plip1 recombinant protein was determined using ExPASy tool and its nucleotide sequence was converted into amino acid sequence. The protein sequence contained 296 amino acids (protein id AHF22385). The deduced amino acid sequence of KBS-plip1 was used to perform a BLAST search of the amino acid sequence database and revealed 55% amino acid sequence homology with uncultured bacterium lipase. The homology modelling was performed using sequence alignment of amino acids of KBS-plip1 against the PDB. Based on the alignment, a model of KBS-plip1 lipase was built using a bacterial lipase (PDB ID: ABZ9) as template for the model building. Sequence alignment showed that the target sequence (KBS-plip1) and template ABZ9 share 72% identity. The ribbon modelled enzyme is a folded dimmer and consists of 9 β-strands attached with 7 α-helices. The enzyme has a catalytic triad, the elements of which are borne on loops, Chapter V: Discussion Page 217, which are the best conserved structural features of the fold. According to homology study, KBS-plip1 was found to be folded dimmer.

The Ramachandran plot for KBS-plip1 predicted protein structure in the region of possible angle formations by φ and ψ. The conventional terms represented the torsion angles on either side of α-carbon in the peptides. As per three regions of Ramachandran plot, favoured, allowed, and outlier were considered and showed 95.6, 4, and 0.4% amino acid residues, respectively. The same was significant as high percentage of residues (>95%) were found in the favoured region. This observation was also supported by Byun et al. (2007) who reported the crystal structure of hyperthermophillic esterase *Est*E1 enzyme.

## REFERENCES

Beisson, F.; et al. Methods for Lipase Detection and Assay: A Critical Review. *Eur. J. Lipid Sci. Technol.* **2000**, *102*, 133–153.

Byun et al. Crystal Structure of Hyperthermophilic Esterase EstE1 and the Relationship Between its Dimerization and Thermostability Properties. *BMC Struct. Biol.* **2007**, *7*, 47–57.

Clark, J. M. Novel Non-template Nucleotide Addition Reactions Catalyzed by Prokaryotic and Eucaryotic DNA Polymerases. *Nucleic Acids Res.* **1988**, *16*, 9677–9686.

Daniel, R. The Metagenomics of Soil. *Nat. Microbiol.* **2005**, *3*, 470–478.

Dugan, K. A.; et al. An Improved Method for Post-PCR Purification for mtDNA Sequence Analysis. *J. Forensic Sci.* **2002**, *47*, 811–818.

Girija, D.; et al. Analysis of Cow Dung Microbiota—A Metagenomic Approach. *Indian J. Biotechnol.* **2013**, *12*, 372–378.

Glogauer, A.; et al. Identification and Characterization of a New True Lipase Isolated Through Metagenomic Approach. *Microb. Cell Fact.* **2011**, *10*, 54.

Hochuli, E.; et al. Genetic Approach to Facilitate Purification of Recombinant Proteins with a Novel Metal Chelate Adsorbent, *Nat. Biotechnol.* **1988**, *6*, 1321–1325.

Jaeger, K. E.; Dijkstra B. W.; Reetz, M. T. Bacterial Biocatalysts: Molecular Biology, Three Dimensional Structures, and Biotechnological Applications of Lipases. *Annu. Rev. Microbiol.* **1999**, *53*, 315–351.

Jiang, X.; et al. Identification and Characterization of Novel Esterases from a Deep-sea Sediment Metagenome. *Arc. Miocrobiol.* **2012**, *194*, 207–214.

Laemmli, U. K. Cleavage of Structural Proteins During the Assembly of the Head of Bacteriophage T4. *Nature* **1970**, *227*, 680–685.

Langin, D.; et al. Gene Organization and Primary Structure of Human Hormonesensitive Lipases: Possible Significance of a Sequence Homology with a Lipase of Moraxella TA144, an Antarctic Bacterium. *Proc. Natl. Acad. Sci.* **1993,** *90,* 4897–4901.

Martin-Laurent, F.; et al. DNA Extraction from Soils: Old Bias for New Microbial Diversity Analysis Methods. *Appl. Environ. Microbiol.* **2001,** *67,* 2354–2359.

Pushpam, P. L.; Rajesh, T.; Gunasekaran, P. Identification and Characterization of Alkaline Serine Protease from Goat Skin Surface Metagenome. *AMB Express.* **2011,** *1,* 1–10.

Robe, P.; et al. Extraction of DNA from Soil. *Eur. J. Soil Biol.* **2003,** *39,* 183–190.

Rondon, M. R.; et al. Cloning and Soil Metagenome: A Strategy for Accessing Thegenetic and Functional Diversity of Uncultured Microorganisms. *Appl. Environ. Microbiol.* **2000,** *66,* 2541–2547.

Sagar, K.; et al. Isolation of Lipolytic Bacteria from Waste Contaminated Soil: A Study with Regard to Process Optimization for Lipase. *Int. J. Sci. Technol. Res.* **2013,** *2,* 214–218.

Schloss, P. D.; Handelsman, J. Biotechnological Prospects from Metagenomics. *Curr. Opin. Biotechnol.* **2003,** *14,* 303–310.

Torsvik, V.; et al. Novel Techniques for Analysing Microbial Diversity in Natural and Perturbed Environments. *J. Biotech.* **1998,** *64,* 53–62.

Urban, M.; Adamczak, M. Exploration of Metagenomes for New Enzymes Useful in Food Biotechnology—A Review. *Pol. J. Food Nutr. Sci.* **2008,** *58,* 11–22.

# CHAPTER 8

# BIOCHEMICAL CHARACTERIZATION OF PURIFIED LIPASE

## CONTENTS

## 8.1  DOSE-DEPENDENT ENZYME ACTIVITY

To determine the effect of enzyme concentration on the catalytic activity, a graded amount of enzyme (0.1–2.0 µg) is added to the reaction mixture and the lipase activity is assayed by the titrametric method.

The dose dependent study demonstrated increased lipolytic activity linearly up to the enzyme concentration of 1.0 $\mu g \cdot ml^{-1}$; however, beyond this concentration, saturation in enzyme activity was observed. The experiment was carried out by graded concentration (0.1–2.0 $\mu g \cdot ml^{-1}$) of enzyme (Fig. 8.1).

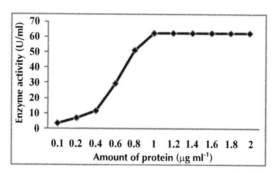

**FIGURE 8.1**   Dose-dependent catalytic activity of lipase.

## 8.2   SUBSTRATE SPECIFICITY AND EFFECT OF SUBSTRATE CONCENTRATION

To determine the substrate specificity of purified KBS-plip1, the different substrates such as tributyrin, olive oil, soybean oil, sunflower oil, palm oil, coconut oil and castor oil are used at a final concentration of 1% (v/v) and incubated at 37°C for 30 min. The lipolytic activity is assayed by titra-metric method. Enzyme assay was done following the same procedure. The recombinant protein showed the maximum activity in the presence of tributyrin (Fig. 8.2).

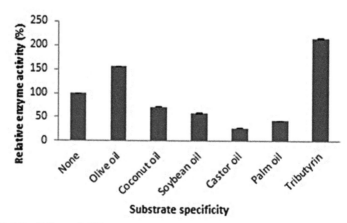

**FIGURE 8.2**   Effect of different substrates on activity of lipase (activity without any additive, 100%).

The effect of substrate concentration on the enzyme activity of KBS-plip1 is determined by titrametric method. Graded concentration of the enzyme substrate ranging from 0.1, 0.2, 0.5, 1.0, 1.5, 2.0, to 2.5% is incubated with 1 μg of purified protein for 15 min at 37°C and then lipase activity assayed (Fig. 8.3). The catalytic activity of lipase showed that with an increase in the tributyrin concentration there was corresponding enhancement of lipid hydrolysis, and the saturation in the enzyme activity was reached at 1.0% tributyrin.

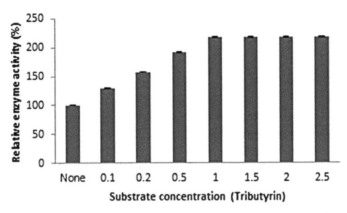

**FIGURE 8.3** Effect of substrate concentration on the activity of lipase (activity without any additive, 100%).

## 8.3 ENZYME KINETICS

### 8.3.1 DETERMINATION OF $K_M$ AND $V_{MAX}$ FOR THE ENZYME CATALYSED REACTIONS

A constant amount of enzyme is incubated with graded concentration 0.1–2.5% of tributyrin emulsion for 10 min at 37°C. The kinetics of catalysed reaction ($K_m$ and $V_{max}$ values) of the purified enzymes is calculated using Lineweaver–Burk plot 350. Lineweaver–Burk plot is drawn using GraphPad Prism 6 software. The enzyme activity is estimated using titrametric method under the standard conditions (Fig. 8.4).

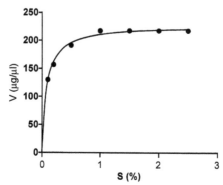

**FIGURE 8.4** Effect of substrate concentration on the initial velocity of lipase catalysed reaction.

The values for $V_{max}$ and $K_m$ were obtained using Lineweaver–Burk plot. Initial velocities of the purified lipase on the different concentrations of tributyrin were determined under the standard assay conditions at pH 7.5. At relatively low concentrations of tributyrin, initial velocity ($V$) increased almost linearly. At substrate concentration of 1%, V increased and beyond this concentration saturation was observed. By plotting the values of 1/v as a function of 1/S, a straight line was obtained that intersect the vertical line at a point which is the $1/V_{max}$ (since $1/[S]=0$, therefore $1/v=1/V_{max}$). Extension of the straight line results in intersecting the horizontal axis (1/[S]) at the point which is $-1/K_m$. Lineweaver–Burk plot was constructed, and $V_{max}$ and $K_m$ were calculated using tributyrin as the substrate. The apparent $V_{max}$ and $K_m$ values of enzyme catalysed reaction are 227 U/mg and 0.0806 mg/ml, respectively (Fig. 8.5).

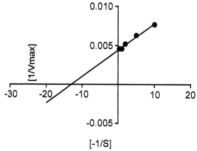

**FIGURE 8.5** Determination of $K_m$ and $V_{max}$ of KBS-plip1 recombinant lipase using Lineweaver-Burk plot. Tributyrin 0.1–2.5% was used as substrate. The Lineweaver-Burk plot was plotted using GraphPad Prism 6 software. The calculated $V_{max}$ and $K_m$ for tributyrin are 227 U/ml and 0.0806 mg/ml, respectively.

In the present investigation, the recombinant protein KBS-plip1 showed maximum residual activity in the presence of 1% tributyrin and the same is also supported by Glogauer et al. (2011). The kinetic parameters were characterized in the presence of the best substrates for KBS-plip1, that is, tributyrin using Michaelis–Menten kinetics model and Lineweaver–Burk plot. The apparent $V_{max}$ and $K_m$ values of enzyme catalysed reaction were determined to be 227 U/mg and 0.0806 mg/ml, respectively. The study has demonstrated lower km value and higher $V_{max}$ value as compared to the earlier reported lipases through metagenomic study, as well as the culturable approach. The lower $K_m$ values indicated the high binding affinity of KBS-plip1 toward its specific substrate tributyrin.

## 8.4   EFFECT OF TEMPERATURE AND PH

To determine the effect of temperature on enzyme activity, a range of 15–65°C is used following the titrimetric method. Each reaction mixture is preincubated for 30 min at the designated temperatures prior to the assay. The optimum temperature for the catalytic activity of lipase was assayed at the temperature range of 10–50°C using tributyrin as the substrate. KBS-plip1 enzyme showed optimum activity at 37°C (Fig. 8.6).

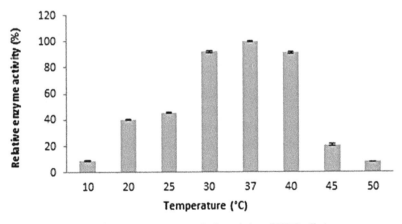

**FIGURE 8.6**   Effect of temperature on catalytic activity of KBS-plip1.

For determining the optimum pH for the catalytic activity of purified KBS-plip1, the titrametric assay is carried out by using buffers of different pH values. These values are obtained as follows 0.1 M sodium acetate buffer (pH 5.5–6.5); 0.1 M phosphate buffer (pH 7.0–7.5); and 0.1 M Tris-HCl buffer (pH 8.0–9.5). The optimum pH for catalytic activity of KBS-plip1 was assayed at the pH range of 3–10 using tributyrin as the substrate. KBS-plip1 showed optimum activity at pH 7.5. The catalytic activity of KBS-plip1 got reduced below and above pH 7.5 (Fig. 8.7).

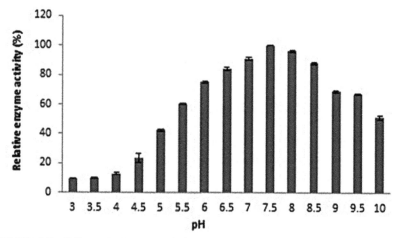

**FIGURE 8.7** Effect of pH on catalytic activity of KBS-plip1. Lipolytic activity of KBS-plip1 at different pH values was assayed using titrimetric method. The various pH values obtained are as follows: 0.1 M sodium acetate, pH 3.0–6.5; 0.1 M potassium $PO_4$, pH 7.0–7.5; and 0.1 M Tris-HCl, pH 8.0–10.0. The maximum measured activity at pH 7.5 was taken as 100%

KBS-plip1 exhibited maximum residual activity at 37°C (Fig. 4.43). At lower temperature KBS-plip1 exhibited relatively high catalytic activity, even at 20°C, the enzyme maintained 40% of its maximum activity. This indicated that KBS-plip1 can be used in the low-temperature organic synthesis of chiral intermediates, which would favour the production of relatively frail compounds. The maximum residual activity of lipases/esterases at 37°C is also supported by Jin et al. (2012). In the present investigation, 7.5 pH was found to be optimum for KBS-plip1. Beyond and above this pH value, there was a decline in enzyme residual activity. The present result has shown the similarity with esterase EstC23 isolated from soil metagenome156. According to this, lipases with alkaline pH have novel stability. The alkaline pH of KBS-plip1 demonstrated its suitability for their incorporation in the commercial laundry detergent formulations, which require the enzyme to remain active at alkaline pH.

## 8.5   EFFECT OF SURFACTANTS

To determine the effect of surfactants, ionic surfactants such as sodium dodecyl sulphate (SDS), nonionic surfactants such as Tween-20, 40, 80, CTAB and Triton X-100 at the final concentration 0.5% are separately added to the reaction mixture and then incubated for 15 min at 37°C, followed by lipase assay with titrimetric method.

To determine the effect of surfactants on the catalytic activity of KBS-plip1 recombinant protein, different surfactants with the final concentration of 0.5% was used. The recombinant protein exhibited the maximum activity in presence of CTAB, while SDS inhibited the lipolytic activity of recombinant protein (Fig. 8.8).

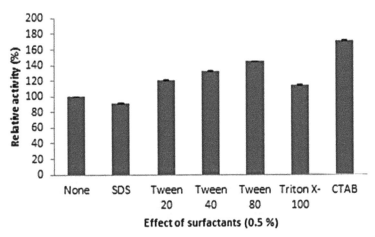

**FIGURE 8.8**   Effect of surfactants on catalytic activity of KBS-plip1. The activity of lipase measured without any additive was considered as 100%.

The effect of surfactants on enzyme activity gives an insight into its nature. According to Rees et al. (2003) the presence of different type of surfactants in any heavy duty laundry detergent, incorporating an enzyme in the detergent becomes a difficult task for the manufacturing industries. To overcome this problem, an effort was made to determine the effect of surfactants on enzyme activity. The purified KBS-plip1 enzyme was found to be stable in the presence of Tween-20, Tween-40, Tween-80, Triton X-100, and CTAB. But the maximum residual activity of KBS-plip1 was observed in the presence of CTAB. SDS inhibited the stability of KBS-plip1.

## 8.6  EFFECT OF OTHER FACTORS

### 8.6.1  METAL IONS

The effect of different divalent cations on lipase activity is assessed titrametrically with the purified protein in the presence of different metal ions like $Li^{2+}$, $Ca^{2+}$, $N^{2+}i$, $Mg^{2+}$, $Hg^{2+}$, $Co,^{2+}$ $Cu^{2+}$, $Fe^{2+}$, Zn, $Mn^{2+}$ and $Cd^{2+}$. The enzyme activity without metal ions serves as the control and is considered as 100%.

To determine the effect of metal ions on the catalytic activity of lipase, the reaction mixture was incubated in the presence of different divalent ions before enzyme assay. KBS-plip1 recombinant lipase showed maximum activity in the presence of $Ca^{2+}$ whereas $Li^{2+}$, $Ni^{2+}$, $Co^{2+}$, $Cu^{2+}$ and $Cd^{2+}$ significantly inhibited the enzyme activity of KBS-plip1 (Fig. 8.9).

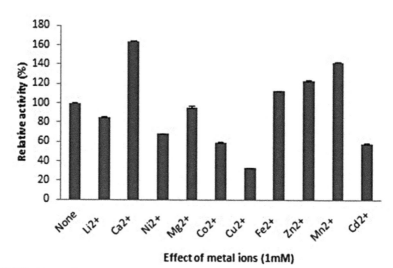

**FIGURE 8.9**  Effect of metal ions on catalytic activity of KBS-plip1. The activity of lipase measured without any additive was considered as 100%.

Most of the enzymes incorporate divalent cations and transition metal ions within their structure to stabilize the folded conformation of the protein. The commercial laundry detergent contains metal chelators as important ingredients to remove metal ions and decrease water hardness. Therefore, alkaline lipases with non-metallic nature are the best for commercial laundry detergent formulations.

The divalent metal ions $Ca^{2+}$, $Fe^{2+}$, $Zn^{2+}$ and $Mn^{2+}$ enhanced the lipase activity, whereas $Li^{2+}$, $Ni^{2+}$, $Mg^{2+}$, $Co^{2+}$, $Cu^{2+}$ and $Cd^{2+}$ inhibited. Most of the lipases reported in the literature are $Ca^{2+}$ and $Mg^{2+}$ dependent. In present investigation, KBS-plip1 was found to be $Ca^{2+}$ dependent but independent of $Mg^{2+}$. Since water hardness depends on the presence of $Ca^{2+}$ and $Mg^{2+}$, KBS-plip1 could be applicable in the laundry detergent formulations.

## 8.6.2 EFFECT OF CHELATING AND EMULSIFYING AGENTS

The role of chelating and emulsifying agent on lipase activity is determined by titrametric assay with the purified protein in the presence of 0.5 mM ethylenediaminetetraacetic acid (EDTA) and 0.5% Gum Arabic separately. The enzyme activity without EDTA and gum arabic serve as the control with 100%.

The catalytic activity of KBS-plip1 lipase was determined in the presence of chelating agent EDTA and emulsifying agent gum arabic. EDTA 1.0 mM inhibited and of gum arabic 0.5% enhanced the enzyme activity (Fig. 8.10 and 8.11).

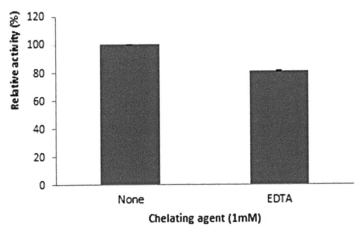

**FIGURE 8.10** Effect of chelating agent on catalytic activity of KBS-plip1. The activity of lipase measured without any additive was considered as 100%.

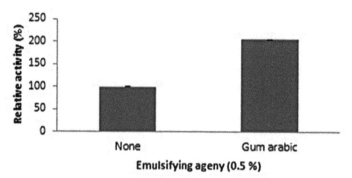

**FIGURE 8.11** Effect of emulsifying agent on catalytic activity of KBS-plip1. The activity of lipase measured without any additive was considered as 100%.

In the present study, EDTA was found to inhibit the catalytic activity of KBS-plip1 enzyme whereas with Gum Arabic it enhanced. The finding is similar to the enzyme reported by Ghori et al. (2011) in the case of lipases isolated from tannery wastes.

### 8.6.3   EFFECT OF COMMON SALT (NACl)

To determine the effect of NaCl on lipase activity, the salt NaCl ranging from 0.1 to 4.0 M is added to the reaction mixture and then the titrametric assay is performed. The enzyme activity without the addition of NaCl is considered as 100%.

Effect of salt on the enzymatic activity of KBS-plip1 was assayed in the presence of NaCl at 0.1–4 M using tributyrin as the substrate. In 1.5 M NaCl, KBS-plip1 exhibited maximum enzymatic activity (Fig. 8.12).

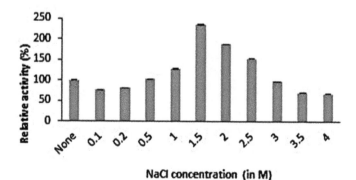

**FIGURE 8.12** Effect of NaCl concentration on catalytic activity of KBS-plip1. The activity of lipase measured without NaCl was taken as 100%.

NaCl at the concentration of 1.5 M enhanced the residual activity of KBS-plip1 by almost 2.5 fold. The catalytic activity of the recombinant KBS-plip1 increased in the range of NaCl concentration 0.5–2.5 M. The stability of KBS-plip1 in the presence of high salt concentration made it to be a good candidate for industrial applications where high salt concentration is used.

### 8.6.4 EFFECT OF ORGANIC SOLVENTS

The stability of the organic solvent of purified lipase is studied by pre-incubating the purified enzyme with the various organic solvents (15 and 30% v/v) viz methanol, ethanol, 1-propanol, 2-propanol, glycerol, acetone, acetonitrile, DMSO, and benzene at 37°C for 120 min. Aliquots are withdrawn at desired time intervals to determine the enzyme activity. The enzyme activity without the organic solvent is considered as 100% activity.

To check stability and activity of KBS-plip1 lipase in the presence of organic solvents, the assay was performed with the different organic solvents. The enzyme exhibited maximum enzymatic activity in 30% of 2-propanol (Fig. 8.13).

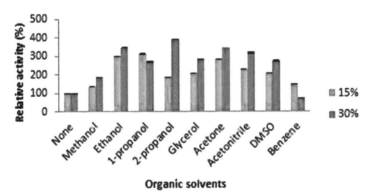

**FIGURE 8.13**   Effect of organic solvents on catalytic activity of KBS-plip1. The activity of lipase measured without any additive was taken as 100%.

The recombinant enzyme, KBS-plip1 has shown the stability in the organic solvents such as acetone, methanol, ethanol, 1-propanol, 2-propanol, glycerol, acetonitrile and DMSO at concentrations of 15 and 30% (v/v). 1-propanol and benzene at 15% (v/v) enhanced the residual activity of KBS-plip1 by three-fold and two-fold, respectively. Acetone, ethanol, methanol, 2-propanol, glycerol, acetonitrile, and DMSO at 30% (v/v) also enhanced the enzyme activity. KBS-plip1 exhibited maximum enzyme activity by 3.5-fold in the presence of 30% 2-propanol (Fig. 4.50). The enhanced activity of KBS-plip1 might be due to the change in the confirmation of the enzyme by organic solvents. The same was reported by Matsumoto et al. KBS-plip1 acted more like a lipase rather than an esterase, because lipases have evolved to tolerate organic solvents more due to the natural pressure of hydrolysing water-insoluble long-chain triglycerides. Ahmed et al. (2010) and Dandavate et al. (2009) reported the similar trend. The elevated enzyme activity and remarkable stability of KBS-plip1 makes it a suitable catalyst for the transesterification reactions.

### 8.6.5   THERMAL STABILITY STUDY

The thermal stability of the KBS-plip1 is investigated by measuring the residual activity of enzyme after incubating at 2.0 mg/ml and temperature ranging from 10 to 100°C for 60 min. The enzyme activity is assayed using titrametric method.

Thermal stability and activity of the KBS-plip1 lipase enzyme were estimated at different temperatures (25–55°C). The enzyme was stable at 37°C for 100 min (Fig. 8.14). However, the stability decreased drastically between 45 and 55°C with half-life of 60 and 20 min, respectively.

Residual activity was measured under standard conditions.

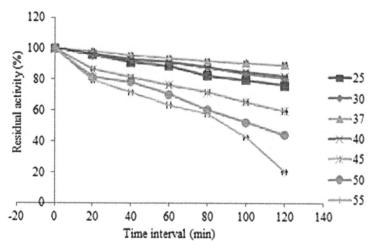

**FIGURE 8.14**   Thermal stability profile of KBS-plip1 purified lipase at different temperatures.

As stated by Saboto et al. (1999) thermostable enzymes are stable and active at temperature that is even higher than the optimum temperature for the growth of the microorganisms. According to Zeikus et al. (1998) thermo-stable enzymes are gaining wide industrial and biotechnological interest due to the fact that these enzymes are better suited for harsh industrial processes. Walsh (2001) also stated that the thermostable enzymes are more stable by inhibiting the denaturizing effect of detergents, organic solvents, chaotropic agents, and oxidizing agents. Thermostable enzymes are resistant to proteolysis. In view of the above important aspect of thermostable enzymes, this study examined the thermostability of recombinant protein. In the present investigation KBS-plip1 exhibited stability or residual activity at 37°C for 100 min with the residual activity greater than 90%, but only 50% residual activity at the temperature range of 38–55°C for 80 min.

The dose dependent study of KBS-plip1 demonstrated that the lipolytic activity increased linearly up to the enzyme concentration of 1.0 $\mu g \cdot ml^{-1}$. The KBS-plip1–tributyrin complex undergoes a lipolytic reaction to release free fatty acids along with the original enzyme. The rate of the chemical reaction is affected by the concentration of enzyme as well as substrate. The rate of enzyme reaction increases with the increase in substrate concentration. At high or low enzyme concentrations, available enzyme active sites could be occupied by the substrates. Therefore, with the increase of the substrate concentration, there would not be further change in the rate of diffusion. In other words, if substrate concentration is constant, with the increase of enzyme there will be increase in lipolytic activity linearly up to enzyme concentration of 1.0 $\mu g \cdot ml^{-1}$ because at this enzyme concentration all active sites of the enzymes are occupied by the substrate. Beyond this enzyme concentration there will be no substrate to bind to the active site of the enzyme.

## REFERENCES

Ahmed, E. H.; Raghavendra, T.; Madamwar, D. An Alkaline Lipase from Organic Solvent Tolerant *Acinetobacter* sp. EH28: Application for Ethyl Caprylatesynthesis. *Bioresour. Technol.* **2010,** *101,* 3628–3634.

Dandavate, V.; et al. Production, Partial Purification and Characterizaing Organic Solvent Tolerant Lipase from *Burkholderia Multivorans* V2 and its Application for Ester Synthesis. *Bioresour. Technol.* **2009,** *100,* 3374–3381.

Ghori, M. I.; Iqbal, M. J.; Hameed, A. Characterization of a Novel Lipase from *Bacillus* sp. Isolated from Tannery Wastes. *Braz. J. Microbiol.* **2011,** *42,* 22–29.

Glogauer, A.; et al. Identification and Characterization of a New True Lipase Isolated Through Metagenomic Approach. *Microb. Cell Fact.* **2011,** *10,* 54.

Jin, P., et al. Overexpression and Characterization of a New Organic Solvent-Tolerent Esterase Derived from Soil Metagenomic DNA. *Bioresour. Technol.* **2012,** *116,* 234–240.

Matsumoto, T.; et al. Yeast Whole-Cell Biocatalyst Constructed by Intracellular Overproduction of Rhizopus Oryzae Lipase is Applicable to Biodiesel Fuel Production. Appl. Microbiol. Biotechnol. **2001,** *57,* 515–520.

Rees, H. C.; et al. Detecting Cellulase and Esterase Enzyme Activities Encoded by Novel Genes Present in Environmental DNA Libraries. *Extremophiles* **2003,** *7,* 415–421.

Saboto, D.; et al. The β-glycosidase from the Hyperthermophilic Archaeon Sulfolobus: Enzyme Activity and Conformational Dynamics at Temperatures above 100°C. *Biophys. Chem.* **1999,** *81,* 23–3.

Walsh, G. *Proteins: Biotechnology and Biochemistry;* John Wiley & sons: England, 2001; pp 189–252.

Zeikus, J.; Vielle, C.; Savachenko, A. Thermozymes: Biotechnology and Structure-Function Relationship. *Extremophiles* **1998,** *2,* 179–183.

# CHAPTER 9

# GENOMIC STUDY OF CULTURE DEPENDENT BACTERIA

## CONTENTS

## 9.1 ISOLATION AND SCREENING FOR LIPASE PRODUCING CULTURABLE BACTERIA

Soil samples collected from the different sources are serially diluted in sterile phosphate buffered saline (PBS) up to $10^{-7}$. From the dilution, 100 µl is plated on Luria Bertani (1951) agar (LBA) plates and incubated overnight at 37°C. Independent pinhead bacterial colonies are collected and liquid cultured in 10 ml sterile tubes containing LB broth. All the pure colonies are screened for lipase production on tributyrin agar (TBA) plates at 37°C for 72 h on the basis of formation of zone of hydrolysis surrounding the bacterial colony.

A total of 16 culturable bacteria were isolated using serial dilution $10^{-1}$–$10^{-6}$ from the bakery soil sample. All the bacterial isolates were screened for lipolytic activity on the TBA medium (Casein enzymic hydrolysate 20.000, bile salts mixture 1.500, Agar 15.000 and pH (at 25°C) 7.2±0.2) containing 1.0% tributyrin. On the basis of formation of zone of hydrolysis, the bacteria were considered as potential lipase-producing bacteria.

## 9.2   PURE CULTURE OF LIPASE-PRODUCING BACTERIAL ISOLATES

Pure cultures of the isolated bacterial strains are established by following the standard protocol. At first, a loopful of bacterial culture is inoculated in nutrient broth (pH 7.0) with 0.5 growths optical density (OD) at 600 nm. After that, 100 µl of culture from the above culture is mixed in 0.9% (w/v) sterile normal saline and serially diluted $10^{-7}$ with a final volume of 2.0 ml. From this dilution, 100 µl aliquot is taken and spread over the sterile nutrient agar plates and kept for 24 h at 37°C in order to obtain single distinct colonies. For obtaining single pure bacterial colony, following the methods are used:

1.  *Spread-plate method*: The individual colonies are separated from the culture of mixed populations of microorganisms, spread on agar plates with a sterile L-shaped glass rod, and then the Petri dish is spun on a turntable. The cells are separated from each other allowing formation of colonies without overlapping.
2.  *Streak-plate method*: The method offers a scope to obtain discrete colonies and pure cultures. One sterile loop or transfer needle is dipped in a suitable diluted suspension of bacterial culture which is streaked on the surface of a solidified agar plate to make a series of parallel, nonoverlapping streak. The plates are incubated overnight at 37°C in inverted position to obtain pure bacterial colonies.

Pure culture of the lipase-producing bacterial isolates was maintained by spread-plate and streak-plate methods on LBA and TBA media. All pure cultures were stored at 4°C till further use. Pure cultures of bacteria are preserved at 4°C in nutrient agar slants for short-term storage and subcultured by transferring them to fresh slants at an interval of 30 days. The isolates are also stored in nutrient broth having 15% (v/v) glycerol and kept at −80°C for long-term storage.

## 9.3   INOCULUM PREPARATION

Inoculum of the isolated bacterial strains is prepared by transferring single colony from a 24 h old culture plate into 5.0 ml of nutrient broth or LB broth and allowed to incubate overnight at 180 rpm and 37°C. The seed culture is used for inoculating the production medium at $10^5$ (v/v). Number of viable cells is presented as CFU/ml by using hemocytometer. The bacterial isolates were recovered, pure cultured and maintained either in stab agar culture at 4°C or preserved in 15% glycerol at −80°C.

## 9.4   ZONE OF HYDROLYSIS

To determine the zone of hydrolysis of lipase-producing bacteria, a clear area surrounding the bacterial growth on TBA plate supplemented with 1% (v/v) tributyrin is measured in millimetre. The measurement of zone of hydrolysis on TBA medium was determined in the presence of 1.0% tributyrin and incubated at 37°C for 37–48 h. All five bacterial isolates showed zone of hydrolysis ≥25 nm surrounding the bacterial colony (Fig. 9.1(a)–(e)).

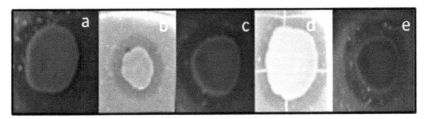

**FIGURE 9.1**   Measurement of zone of hydrolysis of bacterial isolates (a) KBS-101, (b) KB2F, (c) KBS-103, (d) KBS-105 and (e) KBS-107.

A total of 16 culturable bacteria were isolated from bakery waste soil and screened for lipolytic activity on TBA medium. The soil sample collected from the waste deposit site contains an excess amount of carbon which could be a potential source for most the bacteria capable of producing the lipase enzyme. Sagar et al. (2013) reported the isolation of lipolytic bacteria from waste contaminated soil. Bacterial strain KBS-101 showed the largest zone of hydrolysis of 28 mm (Fig. 9.1) followed by KBS-103 (26 mm), KBS-105 (26 mm), KB2F (25 mm) and KBS-107 (25 mm). All five strains were considered to be promising for the production of ester-ases/lipases. Kumar and Gupta (2008) reported lipase/esterase-producing *Bacillus* sp. strain from city garbage using the tributyrin as substrate. On the basis of formation of zone of hydrolysis, five bacterial strains were considered to be promising for the production of lipase and subsequent studies were carried out on these. Mobarak-Qamsari et al. (2011) reported a lipase-producing novel *Pseudomonas aerugenosa* from waste water. The pure culture of lipase-producing bacterial isolates was maintained by spread and streak-plate methods.

## 9.5   GROWTH KINETICS AND LIPASE PRODUCTION

The growth of lipase-producing bacterial strains are observed at 37°C in tributyrin broth media supplemented with 1% (v/v) tributyrin shaking at 200 rpm on a rotary shaker. At the early log phase of bacterial growth, lipase assay of the cell-free culture supernatant is performed using the titrimetric method.

All five bacterial isolates were further characterized based on their growth and lipase yield in submerged fermentation (SmF) (Table 9.1).

**TABLE 9.1**   Growth Kinetics of Lipase-Producing Bacterial Isolates on the Basis of Their Optical Density and Lipase Enzyme Production in Culture Medium.

| Sl No. | Bacterial isolates | Lipase yield (Units/ml) | Growth (OD at 600 nm) |
|--------|--------------------|-------------------------|------------------------|
| 1 | KBS-101 | 125.5 | 0.424 |
| 2 | KB2F | 96.2 | 0.358 |
| 3 | KBS-103 | 112.5 | 0.335 |
| 4 | KBS-105 | 85.8 | 0.453 |
| 5 | KBS-107 | 91.5 | 0.413 |

The next criterion of selection was based on some of the biochemical properties of the culture supernatant containing crude lipase enzyme from the selected bacteria such as stability and compatibility with that of commercial laundry detergents, thermostability and storage stability at 4°C. The culture supernatant containing crude lipase enzyme isolated from the bacterial isolates KBS-101, KB2F, KBS-103, KBS-105 and KBS-107 exhibited superior properties.

Growth kinetics of lipase-producing bacteria was determined with respect to lipase production and OD at 600 nm. Bacteria produce maximum enzyme during the exponential phase because the growth rate of cells gradually increases. This concept was also supported by Bora and Bora (2012). The bacterial strain KBS-101 produced maximum lipase (125.5 U/ml) followed by KBS-103 (112.5 U/ml).

## 9.6  THERMAL STABILITY

The thermal stability of culture supernatant containing crude lipases is investigated by measuring the residual activity after incubating the enzyme at temperature range 30–100°C for 100 min. At every 20 min time interval the enzyme activity is assayed using the titrimetric method.

Thermal stability and activity of the culture supernatant containing crude lipase enzyme from all five bacteria was checked at temperature range of 30–80°C (Fig. 9.2(a)–(e)). The crude enzyme from the bacterial isolates KBS-101, KB2F, KBS-103, KBS-105 and KBS-107 retained 50% of catalytic activity at 60°C for 80 min.

Residual activity was measured under standard conditions. The present study documents the thermostability of KBS-101, KB2F, KBS-103, KBS-105 and KBS-107 and 50% of their catalytic activity retained at 60°C even after 80 min. Krahe et al. (1996) stated that thermostable enzymes facilitate higher reaction rates due to a decrease in viscosity and an increase in diffusion coefficient of substrates and higher process yield due to increased solubility of substrates and products. The thermostability of these enzymes can include them into industrial processes. According to Brockman and Borgstorm (1984), the use of thermostable enzyme not only reduces the cost, but also improves the product quality. Several *Bacillus* sp. were reported to be the main source of lipolytic enzymes, while most of these enzymes are active at a temperature 60°C.

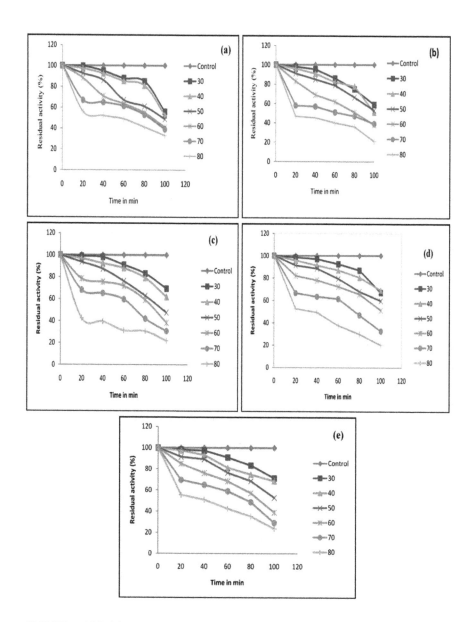

**FIGURE 9.2(a)–(e)**   Thermal stability profile of culture supernatant containing crude lipases from KBS-101, KB2F, KBS-103, KBS-105 and KBS-107 at different temperatures.

## 9.7   EVALUATION OF CRUDE ENZYME IN DETERGENT FORMULATION

The compatibility and stability of culture supernatant containing crude lipase enzymes with some commercial laundry detergents available in the market are accessed. Surf Excel, Rin advance, Wheel (Hindustan Lever Ltd, India), Ghadi (Calcutta detergent Pvt. Ltd. (India) and Tide (Procter and Gamble, India) are examined. Detergents are individually dissolved in tap water at a concentration of 7 mg/ml to stimulate the washing condition, and preheated at 100°C for 60 min to destroy the endogenous protease and lipase activity. The same is reconfirmed by protease and lipase assay. The different concentrations of the culture supernatant containing crude lipase are added to the detergent solution. The relative enzyme activity in the presence of detergent is measured by using the titrimetric method. The activity of crude enzyme with buffer is taken as control with 100% activity.

The suitability, stability and the compatibility of bacterial lipases was checked with other detergents available in the market for the detergent formulation. Lipase from KBS-101, KB2F, KBS-103, KBS-105 and KBS-107 were compatible with Surf Excel, Ariel, Rin advanced, Tide and Wheel. Lipase from bacterial strain KBS-101, KBS-105 and KBS-107 showed maximum enzyme activity in the presence of Surf Excel, and lipase from KB-S102 and KBS-103 showed maximum lipase activity in the presence of Ariel. Similar findings were also reported by Ghanem et al. (2000) in the case of *Bacillus* sp. However, detergent compatibility of all five culture supernatant containing crude lipases demonstrated higher detergent compatibility than the reported alkaline lipases from other bacterial species.

The culture supernatant containing crude lipases from all five bacterial strains were incorporated in the detergent formulations and they exhibited satisfactory catalytic activity under standard washing conditions. Lipase from the bacterial strain KBS-101 and KB2F retained up to 80% of its maximum catalytic activity even after 30 days storage at 4°C; whereas, lipase from the strain KBS-103, KBS-104 and KBS-107 retained 50% of the maximum catalytic activity after 30 days when stored at 4°C.

Detergent ability and stability of the bacterial culture supernatant containing crude lipases was studied to determine the suitability of enzyme for their use in the detergent formulations. Lipase enzyme from KBS-101, KB2F, KBS-103, KBS-105 and KBS-107 were found to be compatible with Surf Excel, Ariel, Rin advanced, Tide and Wheel (Fig. 9.3(a)–(e)).

The enzyme from the bacterial strain KBS-101, KBS-105 and KBS-107 showed maximum activity in the presence of Surf Excel, whereas and KB2F, KBS-103 showed maximum in the presence of Ariel.

## 9.8   STABILITY OF CULTURE SUPERNATANT UNDER STORAGE

The culture supernatant containing crude lipase enzyme is stored for 30 days at 4°C and then its activity determined by following the titrimetric method. The enzyme activity at zero day is considered as 100% activity (Fig. 9.4).

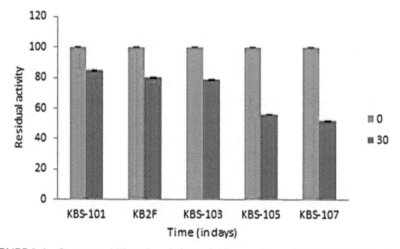

**FIGURE 9.4**   Storage stability of crude bacterial lipases from selected bacterial strains at zero day and post 30 days of storage at 4°C.

The bacterial strains KBS-101 and KB2F retained up to 80%, KBS-103, KBS-104 and KBS-107 50% of their catalytic activity even after 30 days of storage at 4°C.

## 9.9   TAXONOMIC IDENTIFICATION OF
## LIPASE-PRODUCING  BACTERIA

The taxonomic identification of the bacterial isolates is done following the standard morphological, physiological and biochemical tests as described by Cappuccino and Sherman (1992). The bacterial isolates are identified up to genus level with the help of Bergey's manual (Sneath et al., 1986) of Systematic Bacteriology. The identification is based on:

1. Morphological tests
2. Gram's staining
3. Biochemical tests and
4. Ribotyping

## 9.9.1  MORPHOLOGICAL TESTS

The bacterial isolates are characterized by observing their different morphological traits, such as:

1. Size and shape of bacterial colonies
2. Forms of bacterial colonies
3. Margin of bacterial colonies
4. Elevation of bacterial colonies
5. Optical density of bacterial colonies and
6. Different pigmentation of bacterial colonies

The morphology of the isolated and pure cultured bacterial strains was studied in the N-acetyl-muramic acid (NAM) in terms of colony size, pigmentation, form, margin, elevation and Gram stain. The morphology of bacterial isolates and data obtained are presented in Table 9.2.

**TABLE 9.2**   Morphological Characters of Bacterial Isolates Obtained from KBS Soil Sample.

| Sl No | Bacterial isolate | Size | Form | Elevation | Margin | Pigment | Opacity | Gram staining |
|---|---|---|---|---|---|---|---|---|
| 1 | KBS-101 | Large | Circular | Raised | Undulate | White | Opaque | Gram positive |
| 2 | KB2F | Small | Circular | Convex | Entire | White | Translucent | Gram positive |
| 3 | KBS-103 | Small | Circular | Raised | Entire | Cream | Translucent | Gram negative |
| 4 | KBS-105 | Medium | Circular | Raised | Entire | White | Opaque | Gram positive |
| 5 | KBS-107 | Medium | Circular | Flat | Entire | White | Opaque | Gram negative |

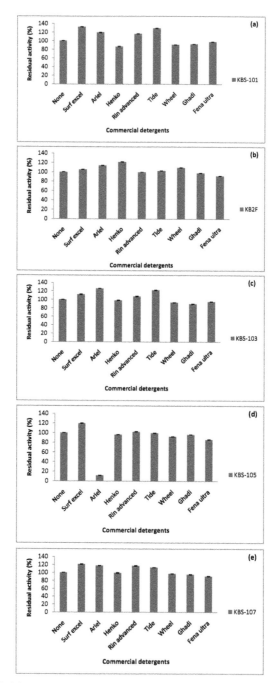

**FIGURE 9.3(a)–(e)** Effect of commercial detergents on compatibility and stability of crude lipases from KBS-101, KB2F, KBS-103, KBS-105 and KBS-107. The activity of lipase measured without any additive was taken as 100%.

## 9.9.2  GRAM'S STAINING

### 9.9.2.1  REAGENTS

**Crystal violet:** Crystal violet 2.0 g is dissolved in 20.0 ml of 95% ethanol and the volume made up to 100.0 ml with distilled water.

**Gram's iodine:** Iodine 1.0 g is dissolved in 300 ml potassium iodide solution.

**Decolouriser:** Absolute alcohol (95% ethyl alcohol) is used as the decoloniser.

**Safranin stain:** Safranin 2.0 g is dissolved in 20.0 ml of 95% ethyl alcohol and the volume made up to 100.0 ml with distilled water.

### 9.9.2.2 PROTOCOL

A thin smear of the 24 h old bacterial culture is prepared on a clean glass slide and kept for 5–10 min for drying followed by heat fixation. Few drops of crystal violet are flooded over the bacterial culture and allowed to stand for 1 min. The excess stain is washed off with distilled water. Gram's iodine is poured over it after primary staining and kept at room temperature for 1 min. The smear is rinsed by washing under tap water followed by addition of a few drops of decolouriser to remove the excess stain. It is washed again with tap water to remove the decolouriser followed by addition of counter stain Safranin for 45 s. The counter stain is removed with tap water and the slide is kept at room temperature for air-drying. The bacteria are observed under a light microscope (100×).

## 9.9.3  BIOCHEMICAL CHARACTERIZATION

The biochemical tests that are conducted for characterization of the bacterial isolates are given in subsequent text.

### 9.9.3.1  CATALASE TEST

The test is performed to determine the presence of catalase enzyme in the bacterial isolates. Trypticase soy agar (TSA, 15 g tryptone, 5 g soytone–enzymatic digest of soybean meal, 5 g sodium chloride and 15 g agar) 10 ml is streaked with a loopful of 24 h bacterial culture and kept at 37°C for 48 h. Hydrogen peroxide ($H_2O_2$) 3% (v/v) is added to detect the production of catalase by the bacterial isolates. The formation of bubble confirms the catalase positive test (production of catalase by bacteria).

### 9.9.3.2   UREASE TEST

To detect the production of urease enzyme by the bacterial isolates, this test is performed. A volume of 0.1 ml of 24 h bacterial culture is inoculated in 10.0 ml of sterile urea broth and kept at 37°C for 48 h. The isolates degrading urea present in the medium have confirmed the production of urease enzyme. The degradation is assessed with the addition of few drops of phenol red into the medium which turns to deep pink due to the urease activity.
    Composition of Urea broth

### 9.9.3.3   CITRATE UTILIZATION TEST

The citrate test is performed to detect the production of citrase enzyme by the bacterial isolates. Sterile Simmons citrate agar 10 ml is prepared and inoculated with 24 h bacterial culture and kept at 37°C for 48 h. Bromothymol blue indicator is added over the surface of the medium. Positive isolates have used citrate as the sole source of carbon, producing citrase enzyme in the medium. In the case of citrate being utilized by bacteria, there is the production of alkaline products with the change of the colour of bromothymol blue from green (at neutral pH 6.9) to blue (at higher pH 7.6) in the medium. The composition of Simmons citrate agar slant medium is presented below:

| | |
|---|---|
| Magnesium sulphate | 0.200 g |
| Ammonium dihydrogen phosphate | 1.000 g |
| Dipotassium phosphate | 1.000 g |
| Sodium citrate | 2.000 g |
| Sodium chloride | 5.000 g |
| Bromothymol blue | 0.080 g |
| Agar | 15.000 g |
| Final pH (at 25°C) | 6.8±0.2 |

### 9.9.3.4   TRIPLE SUGAR IRON (TSI) TEST

The test in tryptic nitrate broth [casein enzymic hydrolysate 20 g, disodium phosphate 2 g, dextrose 1 g, potassium nitrate 1 g, and pH (at 25°C)

7.2±0.2] with three sugars and iron is carried out in order to differentiate bacterial isolates for their ability to ferment glucose, lactose and sucrose or to reduce sulphur to hydrogen sulphide. A loopful of 24 h bacterial culture is streaked over triple sugar iron (TSI) agar slants and kept at 37°C for 48 h. Change of colour of the medium is observed over TSI agar slants. The composition of TSI medium is shown below:

| | |
|---|---|
| Lactose | 1% |
| Sucrose | 1% |
| Glucose | 0.1% |

## 9.9.3.5   NITRATE REDUCTION TEST

The test conducted in the tryptic nitrate medium detects the ability of an organism to reduce nitrate ($NO_3$) to nitrite ($NO_2$) or some other nitrogenous compounds such as molecular nitrogen ($N_2$) using the enzyme nitrate reductase. The nitrate medium contains potassium nitrate as the substrate. A volume of 1.0 ml of 24 h bacterial culture is inoculated in 100.0 ml of sterile trypticase nitrate broth and kept at 37°C for 48 h. To confirm the nitrate reduction capability of the bacterial culture after 48 h of incubation, solution A (Sulphanilic acid), solution B (α-naphthylamine) and trace amounts of zinc powder are mixed with the bacterial cultures. On $NO_3$ being reduced to $NO_2$, the same will react with sulphanilic acid and α-naphthylamine to produce a cherry red colour. If no colour is developed, it indicates either of the following two reactions:

1.  Nitrate is not reduced
2.  Nitrate is reduced even further to compounds other than nitrite ($NH_2$ or $N_2$)

The composition of tryptic nitrate medium is composed of:

## 9.9.3.6   INDOLE PRODUCTION TEST

The indole test is used to identify bacteria capable of producing indole by using the enzyme tryptophanase. Sterile sulphide indole motility (SIM) agar deep tubes are streaked with 24 h bacterial culture over agar surface and kept at 37°C for 48 h. The by-product indole is identified by this test on addition of Kovac's reagent containing HCl, dimethylamino-benzaldehyde and amyl alcohol. Formation of a red layer has referred to the presence of indole.

### 9.9.3.7   H2S PRODUCTION TEST

This test is used to identify those bacteria capable of reducing sulphur. In this test, a loopful of 24 h bacterial culture is inoculated in 10.0 ml of SIM medium agar slants and kept at 37°C for 48 h. The medium contains cysteine, an amino acid containing sulphur and sodium thiosulphate with peptonized iron or ferrous sulphate. Formation of black precipitate indicated the production of hydrogen sulphide ($H_2S$) by the anaerobic respiration by the bacterial cultures.

### 9.9.3.8   LITMUS MILK REACTION TEST

This test is used to identify bacteria capable of transforming the different milk substrates enzymatically into varied metabolic end products. Sterile litmus milk broth 10.0 ml is inoculated with 0.1 ml of 24-h bacterial culture and kept at 37°C for 48 h. After 48 h of incubation, the change in the colour of the medium refers to lactose fermentation, gas formation, curd formation, litmus reduction, peptonization and alkaline reaction. The composition of litmus milk medium:

### 9.9.3.9   METHYL RED-VOGES-PROSKAUER (MR-VP) TEST

1. Methyl red (MR) test is carried out to test the ability of an organism to produce and maintain stable acid end products from the glucose fermentation. In this test, 100.0 ml of sterile methyl red-Voges-Proskauer (MR-VP) broth is inoculated with 1.0 ml of 24 h bacterial culture and kept at 37°C for 48 h. MR-VP medium is divided into two parts A and B. In A, the pH indicator MR is added for confirmation of MR test which detects the presence of large concentrations of acid end products. The tube turning red in colour indicates a positive test and yellow colour indicates a negative result.

2. In part B, mixture of Barritt A and B solutions is added in VP broth. The VP test is used to determine the capability of organisms to produce nonacidic or neutral end products, such as

acetylmethylcarbinol (acetoin), a neutral product formed from pyruvic acid in the course of glucose fermentation. In the VP test, Barritt's reagent consisting of mixture of alcoholic α-naphthol and 40% potassium hydroxide is used. After 15 min of addition of Barritt's reagent, a deep rose colour in the culture medium is developed indicating the presence of acetoin that represents positive result. The absence of rose colouration is a negative result. The composition of MR-VP broth:

| | | |
|---|---|---|
| Glucose | | 5 g |
| Peptone | 5 g | |
| Dipotassium hydrogen phosphate | | 5 g |
| pH | | 6.9 |

## 9.9.3.10   CARBOHYDRATE FERMENTATION MEDIUM

This biochemical test is performed to determine the ability of microorganisms to degrade and ferment carbohydrates with the production of acid and/or gas. For this purpose, the different types of carbohydrate enriched media are used. The composition of phenol red lactose, dextrose and sucrose broths (in $gl^{-1}$) are shown in Table 9.3.

**TABLE 9.3**   Composition of Phenol Red Lactose, Dextrose and Sucrose Broths (in $gl^{-1}$).

| Chemicals | Phenol red, lactose | Phenol red, dextrose | Phenol red, sucrose | Phenol red, xylose | Phenol red, Mannitol |
|---|---|---|---|---|---|
| Protease | 10.0 | 10.0 | 10.0 | 10.0 | 10.0 |
| Peptone | – | – | – | – | – |
| Beef extract | 1.0 | 1.0 | 1.0 | 1.0 | 1.0 |
| Sodium chloride | 5.0 | 5.0 | 5.0 | 5.0 | 5.0 |
| Phenol red | 0.018 | 0.018 | 0.018 | 0.018 | 0.018 |
| Lactose | 10.0 | – | – | – | – |
| Dextrose | – | 1.0 | – | – | – |
| Sucrose | – | – | 5.0 | – | – |
| Xylose | – | – | – | 5.0 | – |
| Mannitol | – | – | – | – | 5.0 |
| pH | 7.4 | 6.9 | 6.9 | 6.8 | 6.8 |

Fermented carbohydrates with the production of acidic wastes might cause yellowing of phenol from red colour, thereby indicating a positive reaction. In some cases, acid production is accompanied by the evolution of $CO_2$ gas which becomes visible as bubbles in the inverted tube.

## 9.10   EXTRACELLULAR ENZYME ACTIVITY

To determine the ability of microorganisms to secrete extracellular hydro-lytic enzymes capable of degrading polysaccharides, lipids and proteins (casein and gelatin), the hydrolysis test is carried out in the concerned medium. The test is used to differentiate microbes based on their ability to hydrolyse starch with exoenzyme amylase.

1.  Tributyrin agar is used to determine the hydrolytic activity of lipase as produced by the tested microorganisms. The medium is composed of nutrient agar supplemented with triglyceride tribu-tyrin as the lipid source. In the experiment, milk agar is used to determine the hydrolytic activity of the enzyme. The positive bacterial isolates produce a clear zone surrounding their growth area in the culture medium. Whereas, in the absence of protease activity, the medium surrounding the growth of the bacteria remains opaque confirming a negative reaction.
2.  Cellulose is a polysaccharide of glucose units in long linear chain linked together by β-1,4 glycosidic bonds. Degradation of cellu-lose is brought about by microbes by the action of extracellular cellulases. Microbial utilization of cellulose was detected using hexadecyltrimethyl ammonium bromide. This reagent precipi-tates intact carboxymethyl cellulose (CMC) in the medium and thus a clear zone around the colony is seen in an otherwise opaque medium indicating degradation of CMC.
3.  Gelatin liquefaction test is used to determine the ability of bacteria to produce hydrolytic exoenzymes called gelatinases that digest and liquefy gelatin. The presence of these enzymes, as determined by the liquefaction, is used for identifying certain bacteria. The composition of Starch agar medium:

Different biochemical tests were performed for all five bacterial strains by following the standard protocols. Data thus obtained are presented in

Table 9.4 Characters recorded for the tests were starch, gelatin, casein, cellulose and lipid hydrolysis, catalase, urease, TSI agar test, citrate utilization, nitrate reduction, indole and $H_2S$ production, litmus milk reaction, MR-VP and glucose utilization and acid production from sugars such as lactose, dextrose, mannitol, sucrose and xylose. Morphology and the results of the biochemical tests stated that isolates KBS-101, KB2F and KBS-105 to belong to genus *Bacillus,* whereas KBS-103 and KBS-104 belong to the genus *Enterobacter.*

**TABLE 9.4** Biochemical Characterization of Lipase-Producing Bacterial Strains.

| Enzymes | Biochemical characterization-screening | | | | |
|---|---|---|---|---|---|
| | **KBS-101** | **KB2F** | **KBS-103** | **KBS-105** | **KBS-107** |
| Shape | Rod | Rod | Rod | Rod | Rod |
| Starch hydrolysis | + | + | + | + | + |
| Gelatin hydrolysis | + | + | − | + | + |
| Casein hydrolysis | + | + | + | + | + |
| Lipid hydrolysis | + | + | + | + | + |
| Cellulose hydrolysis | − | − | + | + | + |
| Catalase | + | + | + | + | + |
| Oxidase | + | + | − | + | + |
| Urease | − | + | + | − | + |
| Citrate | − | − | + | + | + |
| Triple sugar iron | − | + | + | + | + |
| Nitrate reduction | + | + | + | − | + |
| Indole production | − | − | − | | + |
| $H_2S$ production | + | − | − | + | − |
| Litmus milk reaction | Peptoniza-tion | Peptoniza-tion | Peptoniza-tion | Peptoniza-tion | Peptoniza-tion |
| Voges proskauer | + | − | + | + | + |
| Motility | + | − | + | − | + |
| *Acid production from sugars* | | | | | |
| Lactose | − | − | − | − | − |
| Sucrose | + | + | − | + | − |
| Dextrose | + | − | + | + | + |
| Mannitol | + | − | + | − | + |
| Xylose | + | + | + | + | + |

As stated by Prakash et al. (2007) microorganisms are tremendously diverse, both genetically and phenotypically. A polyphasic approach, a unique system for taxonomic characterization of microorganism was applied to characterize lipase-producing bacterial strains, which is based on studying their morphological and biochemical techniques. The isolated and screened five culturable bacterial strains having lipase-producing ability were characterized morphological and biochemical screening. The similar approach for morphological and biochemical characterization was also reported by Cohan et al. (2007) to characterize the bacterial strains. The size of the colonies was observed to fall in the categories of large, moderate, small and pinhead. In most of the cases, white colonies were seen, however, bacterial strain KBS-103 was cream in colour. The difference in morphology existed among bacterial community present in soil environment and also due to the complexity in soil physiological factors such as pH, temperature, soil type, moisture and humic substances could potentially influence bacterial diversity. This was also supported by Robe et al. (2003). These 5 bacterial isolates were subjected to 20 phenotypic tests to provide more descriptive information that would help in recognizing their genus.

Biochemical characterization of the Gram-positive, rod-shaped bacterial strain KBS-101 (KF514427) exhibited positive results for catalase, nitrate reduction, citrate utilization, $H_2S$ production, motility test and VP test. The strain showed negative activity for cellulase, urease and indole production. The strain KBS-101 possessed the ability to hydrolyse starch, gelatine, casein and lipid; also the ability to peptonize the milk protein; and utilize TSI. This strain could produce acid from dextrose, sucrose, xylose, mannitol, but not from lactose. Pradhan et al. (2013) reported the similar data for plant growth promoting *B. methylotrophicus* novel species from rice rhizosphere soil. The Gram-positive, rod-shaped strain KB2F (KF493767) was found to be positive for catalase, urease and oxidase. The strain reduced nitrate to nitrite. Strain KB2F exhibited negative results for cellulase, TSI agar test, indole production, $H_2S$ production, citrate utilization, VP test and motility. Biochemical test for KB2F confirmed its ability to produce acid from sucrose and xylose, but not from lactose, dextrose and mannitol. The bacterium hydrolysed starch, gelatine, casein and lipid, and peptonized the milk. Lesuisse et al. (1993) reported the similar data for *B. subtilis* isolated from Taptapani hotspring, Odisha. The Gram-negative, short, rod-shaped, motile bacterial strain KBS-103 (KF493768) was found to be positive for catalase, urease, VP and citrate utilization. This strain exhibited negative results for the production of indole and $H_2S$. The strain KBS-103 possessed the ability to hydrolyse starch, casein and lipid but not gelatin; it also has the ability to peptonize the milk protein, and utilize TSI. The strain KBS-103 reduced nitrate to nitrite. It produces acid from lactose, sucrose, dextrose, mannitol and xylose. The Gram-positive, rod-shaped, nonmotile bacterial strain KBS-105 (KF493769) exhibited positive results for catalase, cellulase, oxidase, citrate utilization, $H_2S$ production and VP. The strain showed negative activity for urease and indole production. The strain KBS-105 possessed the ability to hydro-lyse starch, gelatine, casein and lipid and the ability to peptonize the milk protein and utilize TSI. The strain reduced nitrate to nitrite.

The biochemical test showed that the strain KBS-105 could produce acid from dextrose, sucrose and xylose, but not from lactose and mannitol. The Gram-negative, rod-shaped, motile strain KBS-107 (KF514428) was found to be positive for catalase, urease, oxidase, cellulase, VP and indole production. Strain KBS-107 exhibited negative results for $H_2S$ production and citrate utilization. The strain reduced nitrate to nitrite. Biochemical test for KB2F confirmed its ability to produce acid from dextrose, mannitol and xylose, but not from lactose and sucrose. The bacterium hydrolysed starch, gelatine, casein and lipid and also had the ability to peptonize the milk protein and utilize TSI. Sadowsky et al. (1983) reported the similar data for *Rhizobium* sp. isolated from soil.

## 9.11 SPECTROPHOTOMETRIC GROWTH DETERMINATION

Absorbance or OD measured through formation of turbidity may not be the actual measure of bacterial number, but the increase in turbidity does indicate bacterial measurement using a photoelectric colorimeter. With the increase in bacterial numbers, the broth becomes more turbid causing the light to scatter, thereby, allowing less light to reach the photoelectric cell. The change in the direction of light is indicated in the spectrophotom-eter as percentage of transmission (%T) and absorbance. Determination of bacterial growth, therefore, requires measurement of samples of a 24 h shaking flask culture to check population size at regular time intervals during the incubation period. Bacterial growth curve often shows four distinct phases of growth: lag (logarithmic), exponential, stationary and death; the actual length of each phase varies with the organism and also the environmental conditions.

### 9.11.1 PROTOCOL

Sterilized LB medium 5.0 ml is taken in different test tubes. Inoculation is done in the test tubes with lipase-producing bacterial strains after proper labelling of the tubes. The cultures are left overnight at 37°C, 180 rpm in the orbital shaker incubator. From each bacterial culture 1.0 ml is inoculated in conical flasks containing 100 ml sterilized LB medium. Inoculated culture flasks are kept in a shaker at 180 rpm and temperatures 25, 30, 37, 40 and 45°C. The absorbance is checked at every 3-h interval by aseptically removing aliquots.

A time and temperature dependence study of the lipase-producing bacterial isolates was conducted. All lipase positive bacterial strains were grown in LB medium at 25–45°C and their growth was monitored spectrophotometrically by taking OD at 600 nm at every 8-h interval. The bacterial strains KBS-101 showed maximum growth at 37°C after 16 h, KB2F at 45°C after 24 h, KBS-103 at 37°C after 16 h, KBS-105 at 37°C after 24 h and KBS-107 at 37°C after 24 h of incubation. The growth pattern of lipase-producing bacterial strains at the different temperatures is presented in Figure 9.5(a)–(e).

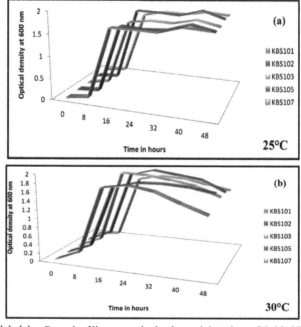

**FIGURE 9.5(a)–(e)** Growth of lipase-producing bacterial strains at 25, 30, 37, 40 and 45°C.

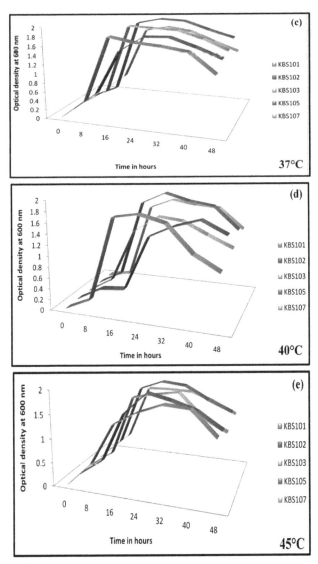

**FIGURE 9.5**   *(Continue)*

The present study documented incubation time and temperature dependent study of lipase-producing bacterial strains. Out of five lipase producing culturable bacterial strains, KBS-101 and KBS-103 were identified to be the most efficient in terms of growth over a temperature range. The similar finding also reported by Hazel (1995) and stated that thermal stresses have strong impacts on the microbial membrane lipids that influence membrane structure and function.

## REFERENCES

Bora, L.; Bora, M. Optimization of Extracellular Thermophilic Highly Alkaline Lipase from Thermophilic *Bacillus sp* Isolated from Hotspring of Arunachal Pradesh, India, *Braz. J. Microbiol.* **2012,** *43,* 30–42.

Brockman, H. L.; Borgstorm, B. *Lipases*; Elsevier: Amsterdam. 3–4, 1984.

Cappuccino, G. J.; Sherman, N. *Microbiology: A Laboratory Manual,* 3rd ed.; Benjamin/cummings Pub. Co.: New York; 1992.

Cohan, F. M.; Perry, E. B. A Systematics for Discovering the Fundamental Units of Bacterial Diversity. *Curr. Biol.* **2007,** *17,* 373–386.

Ghanem, E. H., et al. An Alkalophilic Thermostable Lipase Produced by a New Isolate of *Bacillus alcalophilus. World J. Microbiol. Biotechnol.* **2000,** *16,* 459–464.

Hazel, J. R. Thermal Adaptation in Biological Membranes: is Homeoviscous Adaptation the Explanation? *Annu. Rev. Physiol.* **1995,** *57,* 19–42.

Krahe, M.; Antranikian, G.; Markel, H. Fermentation of Extremophilic Microorganisms. *FEMS Microbiol. Rev.* **1996,** *18,* 271–285.

Kumar, S. S.; Gupta, R. An Extracellular Lipase from *Rhodotorula Mucilaginosa* MSR 54: Medium Optimization and Enantioselective Deacetylation of Phenyl Ethyl Acetate. *Process Biochem.* **2008,** *43,* 1054–1060.

Lesuisse, E.; Schanck, K.; Colson, C. Purification and Preliminary Characterization of the Extracellular Lipase of *Bacillus Subtilis* 168, an Extremely Basic pH-Tolerant Enzyme. *Eur. J. Biochem.* **1993,** *216,* 155–160.

Mobarak-Qamsari, E.; Kasra-Kermanshahi, R.; Moosavi-nejad, Z. Isolation and Identification of a Novel, Lipase-Producing Bacterium, Pseudomonas Aeruginosa KM110. *Iran J. Microbiol.* **2011,** *3*(2), 92–98.

Pradhan, B.; Dash, S. K.; Sahoo, S. *Bacillus Methylotrophicus* sp. nov. A Methanolutilizing, Plant-Growth-Promoting Bacterium Isolated from Rice Rhizosphere Soil. *Asian Pac. J. Trop. Biomed.* **2013,** *3,* 936–941.

Prakash, O., et al. Polyphasic Approach of Bacterial Classification-An Overview of Recent Advances. *Indian J. Microbiol.* **2007,** *47,* 98–108.

Robe, P., et al. Extraction of DNA from Soil. *Eur. J. Soil Biol.* **2003,** *39,* 183–190.

Sadowsky, M. J.; Keyser, H. H.; Bohlool, B. B. Biochemical Characterization of Fast and Slow-Growing Rhizobia that Nodulate Soybeans. *Int. J. Syst. Bacteriol.* **1983,** *33* (4), 716–722.

Sagar, K., et al. Isolation of Lipolytic Bacteria from Waste Contaminated Soil: a Study with Regard to Process Optimization for Lipase. *Int. J. Sci. Technol. Res.* **2013,** *2,* 214–218.

Sneath, P. H. A.; Mair, N. S., Sharpe, M. E.; Holt, J. G. eds. Bergey's manual of systematic bacteriology. Lippincott Williams & Wilkins, Baltimore, USA, 1986; Vol 2.

# CHAPTER 10

# GENOMIC STUDY OF CULTURABLE BACTERIA

## CONTENTS

## 10.1 GENOMIC DNA EXTRACTION OF LIPASE-PRODUCING CULTURABLE BACTERIA

Genomic DNA from bacteria is prepared as described by Sambrook et al. (2001).

### 10.1.1 PROTOCOL

1.  The cells are pelleted by spinning 4 ml of fresh bacterial culture at $9.450 \times g$ for 10 min in a refrigerated Beckman Centrifuge, UK and supernatant decanted.

2.  The supernatant is then discarded, pellet dissolved in 0.8 ml of solution I (50.0 mM glucose, 25.0 mM Tris Cl pH 8.0, 10.0 mM EDTA pH 8.0) and vortexed.

3.  To the resultant mixture, 160.0 µl of lysozyme (10.0 mg·ml$^{-1}$) is added and incubated at room temperature (24°C) for 20 min.

4.  Subsequently, 44.5 µl of 10% (w/v) sodium dodecyl sulphate (SDS) solution is added and reincubated for 10 min at 50°C.

5.  After that, 53.3 µl of RNase (10.0 mg·ml$^{-1}$ stock) is added to the above sample and incubated at 37°C for 90 min. This is followed by addition of 45.3 µl of Na-EDTA (0.1 M, pH 8.0) and reincubated at 50°C for 10 min.

6.  To remove the protein, 26.6 µl of proteinase K (5.0 mg·ml$^{-1}$ stock) is added and incubated at 50°C for 10 min.

7.  Equal volume of phenol (saturated with 0.1 M Tris HCl, pH 8.0) is added to the above solution and mixed thoroughly.

8.  The mixture is centrifuged at 9.450×g for 10 min, the upper (aqueous) phase aspirated into sterile microfuge tube, and lower phase discarded.

9.  Then 700 µl of (1:1) phenol and chloroform-isoamyl alcohol (24:1) are added and mixed thoroughly.

10. Following centrifugation at 9,450 × g for 10 min, the upper phase is transferred to a sterile microfuge tube, then equal volume of chloroform-isoamyl alcohol (24:1) added, and spun at 9,450×g for 10 min. The upper phase is transferred to a sterile microfuge tube and 1/10th volume of sodium acetate (3M, pH 7.0) added to it.

11. The DNA is precipitated by adding 2 volumes of ice-cold absolute ethanol in the above solution, and the DNA pellet recovered by centrifugation.

12. After removal of alcohol, DNA is resuspended in 10 mM Tris HCl–1 mM EDTA buffer (pH 8.0) and stored at 4°C for subsequent uses.

## 10.2  AGAROSE GEL ELECTROPHORESIS

Reagents: Agarose gel, 50X TAE buffer, 1X TAE buffer, loading dye, EtBr.

## 10.2.1  PROTOCOL

- Agarose gel 0.8% is prepared in 1X TAE buffer and heated to dissolve the agarose.
- The same is allowed to cool to around 60°C and then EtBr $(10\,mg \cdot ml^{-1}$ stock) added to make the final concentration of 0.5 g/ml.
- The gel mixture is poured into the gel caster sealed with adhesive tape and fitted with comb.
- The comb and adhesive tape are removed when the gel solidifies.
- The gel is placed in the electrophoresis chamber filled with 1X TAE buffer.
- DNA samples are loaded in gel wells and run at 75 volts/cm till the loading dye reached 75% of the gel.
- The gel is removed from the electrophoresis chamber and examined on ultraviolet (UV) transilluminator.

DNA sample preparation: The DNA samples were mixed with 2.0 μl of loading dye for 5.0–10.0 μl of sample.

## 10.3  QUANTIFICATION AND PURITY DETERMINATION OF THE ISOLATED DNA

After the isolation of DNA, quantification and the quality analysis are necessary to ascertain the approximate quantity of DNA obtained its suitability for further analysis. This is important for the various applications like digestion of DNA by restriction endonuclease enzymes or polymerase chain reaction (PCR) amplification of target DNA. Two commonly used methods for the quantification of nucleic acids are: gel electrophoresis and spectrophotometric analysis was used. In the present investigation the spectrophotometric analysis was used. A ratio between 1.8 and 2.0 denotes that the absorption in the UV range is due to nucleic acids. Ratios lower than 1.8 indicate the presence of proteins and/or other absorbers of UV rays. While, a ratio higher than 2.0 indicates that the samples may be contaminated with chloroform or phenol.

The amount of DNA can be quantified using the formula:

$$\text{DNA concentration } (\mu g/ml) = \frac{\text{O.D.260} \times \text{dilution factor} \times 50\,\mu g/ml}{1000}$$

OD value 1.0 = 50 μl/ml (A260). Double stranded (DS) DNA

### 10.3.1  PROTOCOL

Tris-EDTA (TE) buffer 1.0 μl is taken in a cuvette and the spectrophotom-eter calibrated at 260 and 280 nm. DNA sample 10.0 μl is added along with 990 μl TE buffer and mixed gently. TE buffer is also used as a blank in another cuvette. The OD values at 260 and 280 nm are noted and the ratio is calculated.

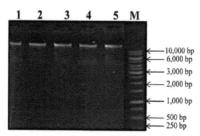

**FIGURE 10.1**   Isolation of genomic DNA from lipase-producing bacteria. Lane M—1 kb DNA ladder (MBI Fermentas, Germany); lane 1—genomic DNA extracted from KBS-101; lane 2—genomic DNA from KB2F; lane 3—genomic DNA from KBS103; lane 4—genomic DNA from KBS105; lane 5—genomic DNA from KBS107.

Genomic DNA from the lipase-producing bacteria was isolated. The isolated DNA was electrophoresed in 0.8% agarose gel. Image of the gel is presented in Figure 10.1. The yield and purity of the isolated DNA was calculated. The purity ratio of the isolated DNA was found to be 1.85–1.94 and yield 33.9–73.4 μg/ml. Data generated are presented in Table 10.1 and Figure 10.2.

**TABLE 10.1**   Purity and Yield of Isolated Genomic DNA.

| Sample` | $A_{260}$ | $A_{280}$ | $A_{260}/A_{280}$ | DNA yield (μg/ml) |
|---------|-----------|-----------|-------------------|-------------------|
| KBS-101 | 1.47  | 0.758 | 1.94 | 73.4 |
| KB2F    | 1.17  | 0.616 | 1.90 | 58.8 |
| KBS-103 | 0.883 | 0.461 | 1.93 | 44.1 |
| KBS-105 | 0.718 | 0.375 | 1.91 | 33.9 |
| KBS-107 | 1.44  | 0.781 | 1.85 | 72.2 |

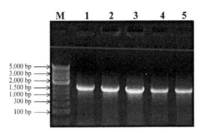

**FIGURE 10.2**  PCR amplification of 16S ribosomal ribonucleic acid (rRNA) gene of lipase-producing bacteria. Lane M—100 bp DNA ladder (Thermo Scientific, USA); lane 1—PCR amplification of KBS-101; lane 2—PCR amplification of KB2F; lane 3—PCR amplification of KBS-103; lane 4—PCR amplification of KBS-105; lane 5—PCR amplification of KBS-107.

The genomic DNA of the lipase-producing culturable bacterial strains was isolated as per the procedure described in Chapter 3. The yield of quality DNA was 73.4 µg/ml, 58.8, 44.1, 33.9 and 72.2 µg/ml in the case of KBS-101, KB2F, KBS-103, KBS-105 and KBS-107, respectively.

## 10.4   MOLECULAR GENETIC ASSESSMENT OF LIPASE-PRODUCING BACTERIA

### 10.4.1   PCR AMPLIFICATION OF 16S RRNA GENE

DNA amplification is performed as per the procedure described 3.4.7. The sequencing of 16S rRNA is presented elsewhere. The standardized PCR condition with respect to each bacterial strain is presented in Table 10.2. The amplified DNA is verified by electrophoresis of aliquots of PCR product (5.0 µl) on a 1.0% agarose gel in 1X TAE (Tris-acetate-EDTA) buffer. The PCR products (16S rRNA gene) are purified using gel extraction kit (Himedia, Mumbai) and sequenced with Big Dye Terminator version 3.1 Cycle sequencing kit and ABI 3500 XL Genetic Analyzer. The sequence data are aligned and compared with already existing sequences obtained from National Center for Biotechnology Information (NCBI) database.

**TABLE 10.2** Optimal Polymerase Chain Reaction (PCR) Conditions for Amplification of Conserved Region of 16S rRNA in the Gene of Selected Lipase-producing Bacterial Strains PCR.

| Conditions | Bacterial strains | | | | |
|---|---|---|---|---|---|
| | **KBS 101** | **KBS 102** | **KBS 103** | **KBS 104** | **KBS 105** |
| Initial denaturation | 94°C for 5 min | 94°C for 5 min | 94°C for 2 min | 94°C for 5 min | 94°C for 2 min |
| Denaturation | 94°C for 1 min | 94°C for 40 s | 94°C for 1 min | 94°C for 1 min | 94°C for 45 s |
| Annealing | 55°C for 2 min | 50°C for 1 min | 53°C for 2 min | 53°C for 2 min | 51°C for 90 s |
| Elongation | 72°C for 3 min | 72°C for 2 min | 72°C for 3 min | 72°C for 3 min | 72°C for 2 min |
| Final extension | 72°C for 7 min | 72°C for 5 min | 72°C for 7 min | 72°C for 7 min | 72°C for 5 min |
| Hol | All at 4°C for 2 min | | | | |
| Cycles | All 35 cycles | | | | |

The composition of the sequencing mix (10.0 μl):

- Big Dye Terminator Ready Reaction Mix: 4.0 μl
- Template (100.0 ng·μl⁻¹): 1.0 μl
- Primer (10 pmol·λ⁻¹): 2.0 μl
- Milli Q Water: 3.0 μl

PCR Conditions for 30 cycles:

- Initial denaturation: 96°C for 1 min
- Denaturation: 96°C for 10 sec
- Hybridization: 50°C for 5 sec
- Elongation: 60°C for 4 min

The PCR amplification of 16S rRNA gene was done using 27F and 1492R universal primers. The size of amplicon of all five lipase-producing bacterial isolates was found to be approximately 1500 bp.

## 10.5 PHYLOGENETIC ANALYSIS

The 16S recombinant DNA (rDNA) sequences of bacteria under study are aligned with reference sequences showing homology from the NCBI database (http://blast.ncbi.nlm.nih.gov) using the multiple sequence alignment programme of MEGA5.2. Phylogenetic trees are constructed using ClustalW by distance matrix analysis and the neighbour joining method (Saitou and Nei, 1987) Phylogenetic trees are displayed using TREEVIEW354. The 16S rRNA gene sequence determined in this study is deposited in GenBank of NCBI data library (http://www.ncbi.nlm.nih.gov/GenBank) under the different accession numbers with respect to each strain.

The 16S rRNA amplified gene of lipase-producing bacterial strain KBS-101was sequenced (Fig. 10.3).

```
>KBS-101
TACGGAATGGGGGCGGGCCCTAATACATGCAAGTCGAGCGGACAGATGGGAGCTTGCTCCCTGATGTTAGCG
GCGGACGGGTGATTAACACGTGGGTAACCTGCCTGTAAGACTGGGATAACTCCGGGAAACCGGGGCTAATAC
CGGATGGTTGTTTGAACCGCATGGTTCAGACATAAAAGGTGGCTTCGGCTACCACTTCCAGATGGACCCGCG
GCGCATTAGCTAGTTGGTGAGGTAACGGCTCACCAAGGCGACGATGCGTAGCCGACCTGAGAGGGTGATCGG
CCACTCTGGGACTGAGACACGGCCCAGACTCCTACGGGAGGCAGCAGTGGGGAATCTTCCGCAATGCAGGAA
AGTTTAAAGGGAGCAACGGCGGGTGCCCCCCCCCCCCGCTCTCCGGAG
```

**FIGURE 10.3** Partial DNA sequences of conserved region of 16S rRNA gene of the lipase-producing bacterial isolate KBS-101.

### 10.5.1 PHYLOGENIC ANALYSIS LIPASE-PRODUCING BACTERIAL STRAIN KBS-101

A homologous search result of the strain KBS-101 demonstrated 95–97% similarity of 16S rRNA sequence was observed with *Bacillus methylotrophicus* strain GSLS13 16S ribosomal RNA gene, partial sequence. The phylogenetic tree constructed from the sequence data by the neighbour joining method showed that *B. methylotrophicus* strain (HQ844459) possessed 97% 16S rRNA sequence homology with KBS-101 and the same represented its closest phylogenetic neighbour. The phylogenetic tree for the strain KBS-101 is shown in Figure 10.4.

The 16S rRNA amplified gene of lipase-producing bacterial strain KB2F was sequenced and homologous search result of the bacterial strain KB2F demonstrated 99% similarity with uncultured bacterium clone AE-C077 esterase/lipase protein gene, complete coding DNA sequence (CDS). As

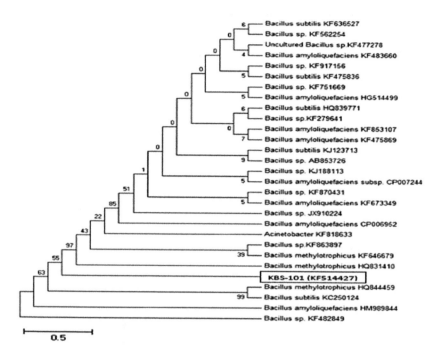

**FIGURE 10.4** Phylogenetic relationship of the strain KBS-101 and other closely related *Bacillus* species based on 16S rDNA sequencing. The data set was resampled 1000 times by using the bootstrap option and percentage values are given at the nodes.

per the phylogenetic tree constructed (neighbour joining method), *Bacillus tequilensis* strain HQ234273 showed 99% 16S rRNA sequence homology with that of KB2F and represented to be its closest phylogenetic neighbour. The phylogenetic tree for the strain KB2F is shown in Figures 10.5 and 10.6.

```
>KB2F
CATTGGGGGCGTGCCTTAATAGTGCAGGTCGAGCGGACAGATGGGAGCTTGCTCCCTGATGTTAGCGGCG
GACGGGTGAGTAACACGTGGGTAACCTGCCTGTGTAAGACTGGGATAACTCCGGGAAACCGGGGCTAATACC
GGATGCTTGTTTGAACCGCATGGTTCAAACATAAAAGGTGGCTTCGGCTACCACTTACAGATGGACCCGC
GGCGCATTAGCTAGTTGGTGAGGTAATGGCTCACCAAGGCAACGATGCGTAGCCGACCTGAGAGGGTGAT
CGGCCACACTGGGACTGAGACACGGCCCAGACTCCTACGGGAGGCAGCAGTAGGGAATCTTCCGCAATGG
ACGAAAGTCTGACGGAGCAACGCCGCGTGAGTGATGAAGGTTTTCGGATCGTAAAGCTCTGTTGTTAGGG
AAGAACAAGTACCGTTCGAATAGGGCGGTACCTTGACGGTACCTAACCAGAAAGCCACGGCTAACTACGT
GCCAGCAGCCGCGGTAATACGTAGGTGGCAAGCGTTGTCCGGAATTATTGGGCGTAAAGGGCTCGCAGGC
GGTTTCTTAAGTCTGATGTGAAAGCCCCCGGCTCAACCGGGGAGGGTCATTGGAAACTGGGGAACTTGAG
TGCAGAAGAGGAGAGTGGAATTCCACGTGTAGCGGTGAAATGCGTAGAGATGTGGAGGAACACCAGTGGC
GAAGGCGACTCTCTGGTCTGTAACTGACGCTGAGGAGCGAAAGCGTGGGGAGCGAACAGG
```

**FIGURE 10.5** Partial DNA sequences of conserved region of 16S rRNA gene of the lipase-producing bacterial isolate KB2F.

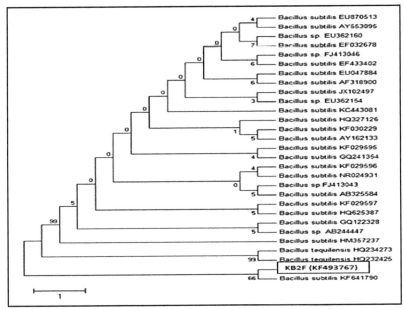

**FIGURE 10.6**  Phylogenetic relationship of the strain KB2F and other closely related *Bacillus* sp. based on 16S rDNA sequencing. The data set was resampled 1000 times by using the bootstrap option and percentage values are given at the nodes.

## 10.5.2   PHYLOGENY OF LIPASE-PRODUCING BACTERIAL STRAIN KBS-103

The 16S rRNA amplified gene of lipase-producing bacterial strain KBS-103 was sequenced and the sequence is shown in Figure 10.7.

```
>KBS-103
CAGGATGGGTATAGATAGATGGTGGGGTAACGGCTCACCTAGGGGACGGATCCCTAGCTGGTTTGGAGAG
GATGGCCAGCCACCCTGGAACTGAGACACGGTCCAGACTCCTACGGGAGGCAGCAGTGGGGGAATATTGCA
CAATGGGCGCAAGCCTGATGCAGCCATGCCGCGTGTATGAAGAAGGCCTTCGGGTTGTAAAGTACTTTCA
GCGGGGAGGAAGGCGTTGAGGTTAATAACCTCAGCGATTGACGTTACCCGCAGAAGAAGCACCGGCTAAC
TCCGTGCCAGCAGCCGCGGTAATACGGAGGGTGCAAGCGTTAATCGGAATTACTGGGCGTAAAGCGCACG
CAGGCGGTCTGTCAAGTCGGATGTGAATCCCCGGGCTCAACCTGGGAACTGCATTCGAAACTGGCAGGCT
AGAGTCTTGTAGAGGGGGGTAGAATTCCAGGTGTAGCGGTGAAATGCGTAGAGATCTGGAGGAATACCGG
TGGCGAAGGCGGCCCCCTGGACAAAGACTGACGCTCAGGTGCGAAAGCGTGGGGAGCAAACAGGATTAGA
TACCCTGGTAGTCCACGCCGTAAACGATGTCGACTTGGAGGTTGTGCCCTTGAGGCGTGGCTTCCGGAGC
TAACGCGTTAAGTCGACCGCCTGGGGAGTACGGCCGCAAGGTTAAAACTCAAATGAATTGACGGGGGCCC
GCACAAGCGGTGGAGCATGTGGTTTAATTCGATGCAACGCGAAGAACCTTACCTACTCTTGACATCCAGA
GAACTTTCCAGAGATGGATTGGTGCCTTCGGGAACTCTGAGACAGGTGCTGCATGGCTGTCGTCAGCTCG
TGTTGTGAAATGTTGGGTTAAGTCCCGCAACGAGCGCAACCCTTATCCTTTGTTGCCAGCGGTCCGGCCG
GGAACTCAAAGGGAGACTGCCAGTGATAAACTGGAGGAAGGTGGGGATGACGTCAAGTCATCATGGCCCTT
ACGAGTAGGGCTACACACGTGCTACAATGGCGCATACAAAGAGAAGCGACCTCGCGAGAGCAAGCGGACC
TCATAAAGTGCGTCGTAGTCCGGATTGGAGTCTGCAACTCGACTCCATGAAGTCGGAATCGCTAGTAATC
GTAGATCAGAATGCTACGGTGAATACGTTCCCGGGCCTTGTACACACCGCCCGTCACACCATGGGAGTGG
GTTGCAAAAGAAGTAGGTAGCTTAACCTTCGGGAGGGCGCTAACCAACTTTGGGTTTCAAGG
```

**FIGURE 10.7**  Partial DNA sequences of conserved region of 16S rRNA gene of the lipase-producing bacterial isolate KBS-103.

A homologous search result of strain KBS-103 demonstrated 98–99% similarity of 16S rRNA sequence with the other species of genus *Enterobacter* and uncultured bacterium. The phylogenetic tree (neighbour joining method) of uncultured organism strain (HQ764973) exhibited 99% 16S rRNA sequence homology with that of KBS-103 and represented to be its closest phylogenetic neighbour (Fig. 10.8).

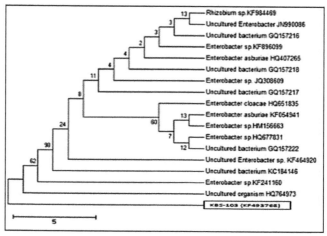

**FIGURE 10.8**    Phylogenetic relationship of the strain KBS-103 and other closely related *Enterobacter* species based on 16S rDNA sequencing. The data set was resampled 1000 using the bootstrap option and percentage values are given at the nodes.

### 10.5.3    PHYLOGENY OF LIPASE-PRODUCING BACTERIAL STRAIN KBS-105

The 16S rRNA amplified gene of lipase-producing bacterial strain KBS-105 was sequenced and the sequence data are shown in Figure 10.9.

```
>KBS-105
CTGGGGGGGTGCCTATACATGCAAGTCGAGCGGACTTGACGGAAGCTTGCTTCCGTTCAAGTTAGCGGC
GGACGGGTGAGTAACACGTGGGTAACCTGCCTGTGTAAGACTGGGATAACTCCGGGAAACCGGGGCTAATAC
CGGATATTCTTTTTCTTCGCATGAAGAAGAATGGAAAGGCGGCTTTTAGCTGTCACTTACAGATGGACCC
GCGGCGCATTAGCTAGTTGGTGAGGTAACGGCTCACCAAGGCAACGATGCGTAGCCGACCTGAGAGGGTG
ATCGGCCACACTGGGACTGAGACACGGCCCAGACTCCTACGGGAGGCAGCAGTAGGGAATCTTCCGCAAT
GGACGAAAGTCTGACGGAGCAACGCCGCGTGAGTGAAGAAGGTTTTCGGATCGTAAAGCTCTGTTGTCAG
GGAAGAACAAGTACGGAAGTAACTGTCCGTACCTTGACGGTACCTGACCAGAAAGCCACGGCTAACTACG
TGCCAGCAGCCGCGGTAATACGTAGGTGGCAAGCGTTGTCCGGAATTATTGGGCGTAAAGCGCGCGCAGG
CGGCTTCTTAAGTCTGATGTGAAAGCCCACGGCTCAACCGTGGGAGGGTCATTGGAAACTGGGAGGCTTGA
GTGCAGAAGAGGAGAGCGGAATTCCACGTGTAGCGGTGAAATGCGTAGAGATGTGGAGGAACACCAGTGG
CGAAGGCGGCTCTCTGGTCTGTAACTGACGCTGAGGCGCGAAAGCGTGGGGAGCGAACAGGATTAGATAC
CCTGGTAGTCCACGCCGTAAACGATGAGTGCT
```

**FIGURE 10.9**    Partial DNA sequences of conserved region of 16S rRNA gene of the lipase-producing bacterial isolate KBS-105.

A homologous search result of the bacterial strain KBS-105 demon-strated 99–100% similarity of 16S rRNA sequence with the other species of the genus *Bacillus*. The phylogenetic tree (neighbour joining method) showed 100% 16S rRNA sequence homology of *Bacillus badius* strain (GQ497939) with KBS-105 and represented to be its closest phyloge-netic neighbour. The phylogenetic tree for the bacterial strain KBS-105 is shown in Figure 10.10.

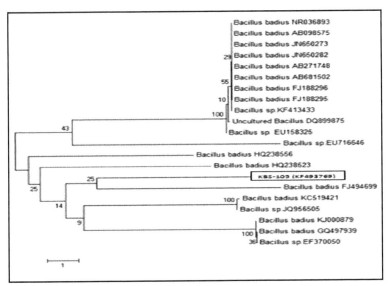

**FIGURE 10.10**   Phylogenetic relationship of strain KBS-105 and other closely related *Bacillus* species based on 16S rDNA sequencing. The data set was resampled 1000 times by using the bootstrap option and percentage values are given at the nodes.

### 10.5.4   *PHYLOGENY OF LIPASE-PRODUCING BACTERIAL STRAINS KBS-107*

The 16S rRNA amplified gene of lipase-producing bacterial strain KBS-107 was sequenced and the data are presented in Figure 10.11.

```
>KBS-107
TGAGATGACAGCCACACTTGGACTGAGACACGTCCAGATCCTACGGAGGCAGCAGTGGGATTTGCACATG
GGCGCAGCCTGATGCAGCATGCGCGTGTATGAGAGCTCGGGTGTAAAGTACTTCAGCGGGAGAGCGTGAG
TATACTCAGCGATGACGTACCGCAGAGAGCACGGCTAACTCGTGCCAGCAGCGCGGTAATACGGAGGGTG
CAAGCGTAATCGGAATTACTGGGCGTAAAGCGCACGCAGGCGGTCTGTCAAGTCGGATGTGAAATCCCCG
GGCTCAACTGGGAACTGCATTCGAAACTGGCAGGCTAGAGTCTTGTAGAGGGGGGTAGAATTCCAGGTGT
AGCGGTGAAATGCGTAGAGATCTGGAGGAATACCGGTGGCGAAGGCGGCCCCTGGACAAAGACTGACGCT
CAGGTGCGAAAGCGTGGGGAGCAAACAGGATTAGATACCCTGGTAGTCCACGCCGTAAACGATGTCGACT
TGGAGGTTGTGCCCTTGAGGCGTGGCTTCCGGAGCTAACGCGTTAAGTCGACCGCCTGGGGAGTACGGCC
GCAAGGTTAAAACTCAAATGAATTGACGGGGGCCCGCACAAGCGGTGGAGCATGTGGTTTAATTCGATGC
AACGCGAAGAACCTTACCTACTCTTGACATCCAGAGAACTTTCCAGAGATGGATTGGTGCCTTCGGGAAC
TCTGAGACAGGTGCTGCATGGCTGTCGTCAGCTCGTGTTGTGAAATGTTGGGTTAAGTCCCGCAACGAGC
GCAACCCTTATCCTTTGTTGCCAGCGGTCCGGCCGGGAACTCAAAGGAGACTGCCAGTGATAAACTGGAG
GAAGGTGGGGATGACGTCAAGTCATCATGGCCCTTACGAGTAGGGCTACACACGTGCTACAATGGCGCAT
ACAAAGAGAAGCGACCTCGCGAGAGCAAGCGGACCTCATAAAGTGCGTCGTAGTCCGGATTGGAGTCTGC
AACTCGACTCCATGAAGTCGGAATCGCTAGTAATCGTAGATCAGAATGCTACGGTGAATACGTTCCCGGG
CCTTGTACACACCGCCCGTCACACCATGGGAGTGGGTTGCTAAAAGAAGTAGGTAGCTTAACCTTCGGGA
GGGCGCTTACTATATGTGTTCCCTCG
```

**FIGURE 10.11**   Partial DNA sequences of conserved region of 16S rRNA gene of the lipase-producing bacterial isolate KBS-107.

A homologous search result of strain KBS-107 demonstrated 95–97% similarity of 16S rRNA sequence was observed with the genus *Entero bacter* and unculturable bacterium. The phylogenetic tree showed 97% 16S rRNA sequence homology of *Enterobacter* sp. (FJ440555) with KBS-107 and represented to be its closest phylogenetic neighbour. The phylogenetic tree for the bacterial strain KBS-107 is shown in Figure 10.12.

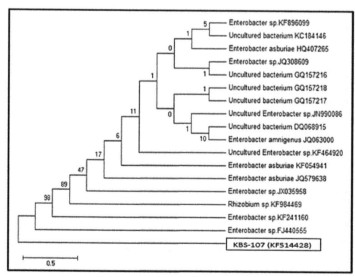

**FIGURE 10.12**   Phylogenetic relationship of the bacterial strain KBS-107 and other closely related bacterial species based on 16S rDNA sequencing. The data set was resampled 1000 times by using the bootstrap option and percentage values are given at the nodes.

Genomic DNA of all five lipase-producing culturable bacteria was processed for PCR amplification using 27F and 1492R universal primers. The size of amplicons was ~1500 bp. The sequencing of PCR amplified products was done using ABI PRISM® 3100 Genetic Analyzer. The homology analysis indicated that the KBS-101 (GenBank entry: KF514427) showed 95–97% 16S rRNA sequence similarity with other species of the genus *Bacillus*. Further, the BLAST (NCBI) search and phylogenetic tree constructed from the sequence data exhibited *B. methylotrophicus* strain (HQ844459) having 97% 16S rRNA sequence homology with KBS-101(GenBank entry: KF514427) and represented it to be the closest phylogenetic neighbour. Homologous search result of the strain KB2F (GenBank entry: KF493767) demonstrated 100% similarity of 16S rRNA sequence with the genus *Bacillus*.

Further, BLAST (NCBI) search showed *Bacillus* sp. (GenBank entry: HQ234273) having 97% 16S rRNA sequence homology with KB2F (GenBank entry: KF493767). The phylogenetic analysis (Fig. 10.12) of KB2F (GenBank entry: KF493767) indicated *Bacillus subtilis* (GenBank entry: KF641790) represented to be its closest phylogenetic neighbour. Homologous search result of strain KBS-103 (GenBank entry: KF493768) demonstrated 98–99% similarity of 16S rRNA sequence with other species of the genus *Enterobacter* and uncultured bacterium. Further, BLAST (NCBI) search showed that *Enterobacter* sp. (GenBank entry: KF896099) and uncultured bacterium (GenBank entry: GQ157218) having 99% 16S rRNA sequence homology with KBS-103 (GenBank entry: KF493768). The phylogenetic tree constructed from the sequence data by the neighbour joining method represented uncultured bacterial strain (HQ764973) to be its closest phylogenetic neighbour. The homology analysis of strain KBS-105 (GenBank entry: KF493769) demonstrated 99–100% similarity of 16S rRNA sequence with other species of the genus *Bacillus*. Further, BLAST (NCBI) search showed *Bacillus badius* strain (GenBank entry: GQ497939) has 100% 16S rRNA sequence homology with KBS-105 (GenBank entry: KF493769). The phylogenetic tree constructed from the sequence data represented *B. badius* strain (GenBank entry: FJ494699) to be its closest phylogenetic neighbour. The results of the study demonstrated that KBS-105 (GenBank entry: KF493769) might belong to *B. badius*.

Homologous search result of strain KBS-107 (GenBank entry: KF514428) demonstrated 96–97% similarity of 16S rRNA sequence with

*Enterobacter* sp. and uncultured bacterium. Further, BLAST (NCBI) search showed *Enterobacter* sp. (GenBank entry: FJ440555) having 97% 16S rRNA sequence homology with KBS-107 (GenBank entry: KF514428). The phylogenetic tree constructed from the sequence data represented that *Enterobacter* sp. (GenBank entry: FJ440555) to be its closest phylogenetic neighbour. The results of this study demonstrated that KBS-107 (GenBank entry: KF514428) might belong to *Enterobacter* sp.

## 10.6 OPTIMIZATION OF CULTURE CONDITION FOR LIPASE PRODUCTION

The following criteria are adopted for screening of parameters influencing lipase yield. As shown in Figure 10.13, with an increase in inoculum size from 0.5 to 2.0% into 100 ml culture medium, the production of lipase by the bacterial strains KBS-101, KB2F, KBS-103, KBS-105 and KBS-107 was enhanced linearly, while increasing from 2.0 to 5.0% ml did not possess much impact in enhancing the lipase production ($p \geq 0.05$). Rather an increase in inoculum size beyond 2% resulted in a steady decline in lipase production by KBS-101, KB2F, KBS-103, KBS-105 and KBS-107.

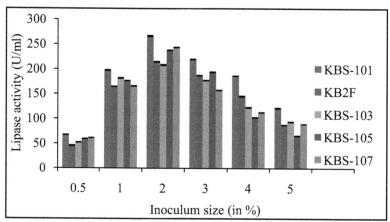

**FIGURE 10.13**  Effect of inoculum size on lipase production from bacterial strains (± S.D of three determinations).

An inoculum size of 2% (v/v; $5 \times 10^6$ cells/ml) produced the highest lipase activity by all five bacterial strains. The activity of the lipases produced by the bacterial strains KBS-101, KB2F, KBS-103, KBS-105 and KBS-107 was 267, 215, 209, 238 and 243 U/ml, respectively. Ito et al. (2001) stated that higher cell density results into higher production of enzyme. But at higher inoculum size above 5%, the enzyme activity reduced to less than 100 U/ml. This could be due to the depletion of oxygen on account of high cell concentration.

## 10.7 EFFECT OF CARBON AND NITROGEN SOURCES FOR THE GROWTH OF CULTURABLE BACTERIA

### 10.7.1 EFFECT OF CARBON SOURCES

Carbon sources separately at a final concentration of 1% (w/v or v/v) (starch, fructose, maltose, lactose, sucrose, galactose, dextrose, mannitol, sorbitol and xylose) is added in the culture medium to check the production of the lipase by selected bacterial isolates.

Various carbon sources were tested for lipase production and the maximum lipase production by the bacterial strains KBS-101 and KBS-105 was observed in the presence of 1.0% glucose followed by starch while KBS-103 and KBS-107 showed maximum lipase production in presence of 1.0% of galactose followed by lactose. The maximum lipase production by the strain KB2F was observed in presence of 1% lactose. Thus data obtained are presented in Figure 10.14(a)–(e).

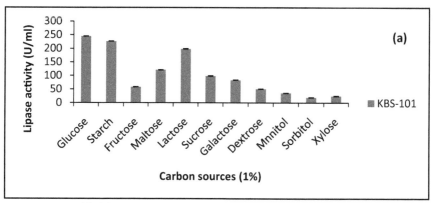

**FIGURE 10.14(a)–(e)**   Effect of the different carbon sources on lipase production from bacterial strains KBS-101, KB2F, KBS-103, KBS-105 and KBS-107.

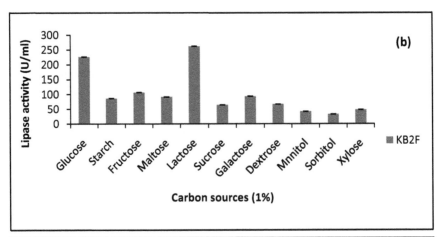

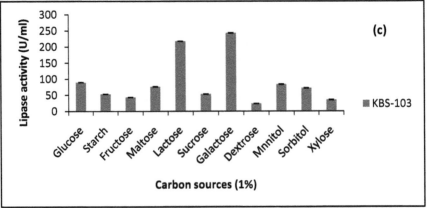

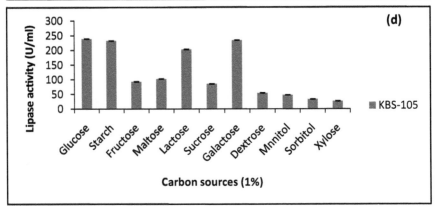

FIGURE 10.14(a)–(e)    *(Continued)*

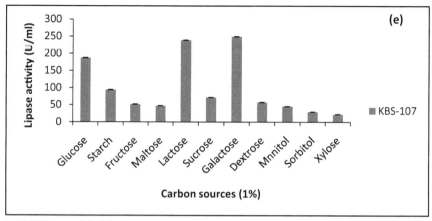

FIGURE 10.14(a)–(e)    *(Continued)*

## 10.7.2   *EFFECT OF VARIOUS NITROGEN SOURCES*

Bacterial isolates are cultured in media supplemented with the various organic and inorganic nitrogen sources at 1% w/v or v/v (beef extract, peptone, yeast extract, tryptone, urea, potassium nitrate, sodium nitrate, ammonium chloride, ammonium sulphate and ammonium nitrate) so as the determine the best source for lipase production. The cell free culture supernatant is used for lipase assay using the titrimetric method.

Various nitrogen sources were tested for lipase production and the maximum lipase production was obtained in the case of the strain KBS-101 and KB2F in the presence of beef extract followed by yeast extract. KBS-103 and KBS-107 showed the maximum lipase production in the presence yeast extract while KBS-105 showed the maximum lipase production in the presence of peptone. Data are presented in Figure 10.15(a)–(e).

Different carbon sources like starch, fructose, maltose, lactose, sucrose, galactose, dextrose, mannitol, sorbitol and xylose at a concentration of 1.0% was used for the culture of bacterial strains. The maximum lipase production (244.7 U/ml) by the strain KBS-101 (KF514427) was observed on use of glucose. In case of bacterial strain KB2F (KF493767), the maximum lipase production (261.9 U/ml) was in the presence lactose. Glucose also enhanced lipase production in the bacterial strain KB2F. Bonala and Mangamoori (2012) reported enhanced lipase production by *B. tequilensis* when supplemented with lactose and starch in the culture medium. Galactose influenced the lipase production by strain KBS-103 and KBS-107, yielded lipase activity 242.5 and 249.1 U/ml, respectively. The highest accumulation of lipase by bacterial strain KBS-105 showed maximum lipase accumulation in presence of glucose with 238.3 U/ml.

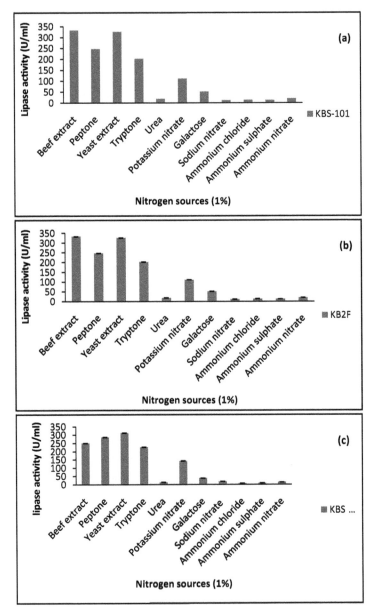

**FIGURE 10.15(a)–(e)**   Effect of the different nitrogen sources on lipase production from bacterial strains KBS101, KB2F, KBS-103, KBS-105 and KBS-107.

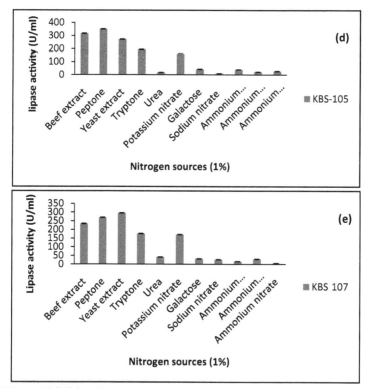

**FIGURE 10.15(a)–(e)** *(Continued)*

Nitrogen sources like beef extract, yeast extract, peptone, tryptone, urea, potassium nitrate, galactose, sodium nitrate, ammonium chloride, ammonium sulphate and ammonium nitrate at the concentration of 1% (w/v) were used for lipase accumulation by the culturable bacterial strains. The study suggested that different bacteria have different preferences for either organic or inorganic nitrogen sources for their growth and lipase production. All five strains showed a preference for both organic and inorganic nitrogen sources for the production of lipase enzyme. Bacterial strains KBS-101, KB2F and KBS-107 showed maximum lipase production 286, 333 and 295 U/ml, respectively in the presence of beef extract, KBS-103 312 U/ml in yeast extract, KBS-105 351 U/ml in peptone. Ananthi et al. (2013) also observed enhanced lipase production in *Bacillus* sp. in the presence of beef extract, yeast extract and peptone as nitrogen sources.

## 10.8   TIME COURSE WITH LIPASE PRODUCTION

To determine the bacterial growth and lipase production in different time intervals, enzyme assay is performed after every 24 h using the titrimetric method. The cell dry biomass and the concentration of enzyme are calculated at 24-h interval.

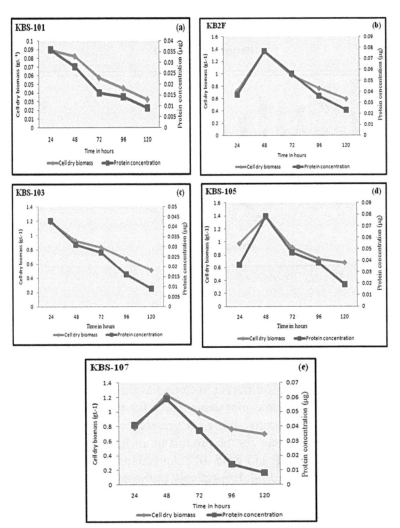

**FIGURE 10.16(a)–(e)**   Time course of lipase production by KBS-101, KB2F, KBS-103, KBS-105 and KBS-107. Legends show bacterial dry biomass and protein concentration of cell free extract.

To determine the bacterial growth in relation to the production of lipase, enzyme assay was performed at every 24 h. The bacterial strains KBS-101 and KBS-103 showed maximum lipase production after 24 h at the cell dry biomass of 0.089 and 1.18 $gl^{-1}$, respectively. The bacterial strains KB2F, KBS-105 and KBS-107 showed maximum lipase production after 48 h at the cell dry biomass of 1.36, 1.39 and 1.23 $gl^{-1}$, respectively. Data thus obtained are presented in Figure 10.16(a)–(e). However, further increase in time after 36 h caused declining bacterial growth and lipase production for all strains.

The time course of lipase production and cell dry biomass yield up to 120 h was assessed in the case of culturable lipase-producing bacterial strains. High lipase production and cell density were obtained in the case of strains KBS-101 and KBS-103 during 24 h; cell dry biomass yield was 0.089 and 1.18 $gl^{-1}$, respectively. Bacterial strain KB2F, KBS-105 and KBS-107 showed maximum lipase production during 48 h with cell dry biomass yielded 1.36, 1.39 and 1.23 $gl^{-1}$, respectively.

## 10.9   EFFECT OF OTHER FACTORS

### 10.9.1   EFFECT OF PH

The pH of the media is adjusted prior to the inoculation of bacteria to determine the effect of pH for lipase production. The cell free culture supernatant is used to determine the lipase activity.

The effect of pH on the production of lipase from the strains KBS-101, KB2F, KBS-103, KBS-105 and KBS-107 was determined by adjusting pH of the culture medium and assaying the enzyme production of the particular pH. Data showed that along with the increase of pH from 7.0 to 9.0, lipase production was enhanced significantly. However, a further increase in the pH beyond 10.0 resulted in the decline of lipase production. Bacterial strain KBS-101 and KBS-105 showed maximum lipase production at pH 8.5; KB2F and KBS-103 at pH 7.5; whereas KBS-107 at pH 9.0. Data obtained are presented in Figure 10.17(a)–(e).

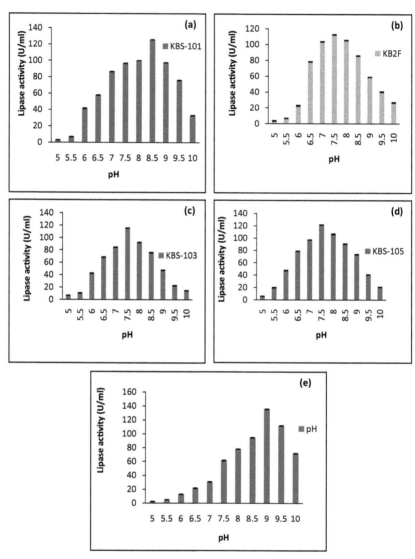

**FIGURE 10.17**　Effect of pH on lipase production from bacterial strains KBS-101, KB2F, KBS-103, KBS-105 and KBS-107.

## 10.9.2　EFFECT OF TEMPERATURE

The influence of temperature on lipase production is determined by incubating the culture flasks in the temperature range at 25–50°C in an orbital shaking incubator. Increase in culture temperature from 25 to 45°C caused significant increase of lipase production by the bacterial strains KBS-101, KB2F, KBS-103, KBS-105 and KBS-107; but beyond 45°C there was a decline. The bacterial strain KBS-101, KBS- 103, KBS-105 and KBS-107 showed maximum lipase production at 37°C, whereas KB2F showed at 45°C. Therefore, 37°C was considered as the optimum temperature for the production of lipase by the bacterial strains KBS-101, KBS-103, KBS-105 and KBS-107, and 45°C was the optimum for lipase production from KB2F. Data obtained are presented in Figure 10.18(a)–(e).

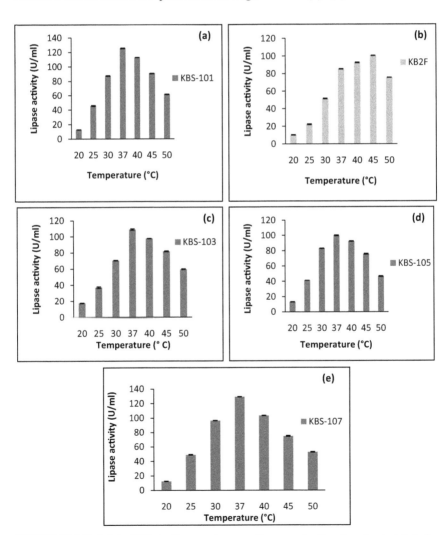

**FIGURE 10.18(a)–(e)**    Effect of temperatures on lipase production from bacterial strains KBS-101, KB2F, KBS-103, KBS-105 and KBS-107.

Lipases from all five bacterial strains were most active at pH 7.0 and 9.0. However, a further increase in pH beyond 10.0 caused a decline in lipase production. In general bacterial lipases are stable in a wide range of pH from 4 to 11. Gupta et al. (2004) reported that maximum activity of lipases at pH values higher than 7.0. Bacterial strain KBS-101 and KBS-105 showed maximum lipase production at pH 8.5, KB2F and KBS-103 at pH 7.5 while KBS-107 at pH 9.0. A temperature range of 25–45°C was found to be suitable for lipase accumulation in all five culturable bacterial strains, but temperature beyond this caused decline in the enzyme accumulation. Parr et al. (1983) reported that most of the soil microorganisms are meso-philes and exhibits maximum growth and enzyme production in the range of 20–35°C. The kinetics of enzymatic reaction for lipase is more conductive at this temperature range. Bacterial strain KBS-101, KBS-103, KBS-105 and KBS-107 exhibited maximum lipase production at 37°C, only KB2F at 45°C. Therefore, 37°C was considered as optimum temperature for maximum lipase accumulation from KBS-101, KBS-103, KBS-105 and KBS-107.

### 10.9.3  EFFECT OF INCUBATION TIME

Influence of incubation time on lipase production is determined by time duration of 6–30 h after the incubation of flasks at the desired tempera-ture. The production of lipase enzyme from the bacterial strains KBS-101, KBS-103, KBS-105 and KBS-107 was maximum after 24 h of incuba-tion at 37°C while KB2F showed after 48 h of incubation at 45°C. Data obtained are presented in Figure 10.19(a)–(e).

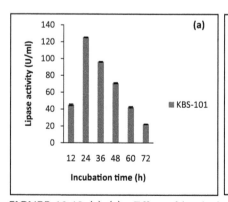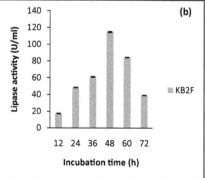

**FIGURE 10.19 (a)–(e)**   Effect of incubation time on lipase production from bacterial strains KBS-101, KB2F, KBS-103, KBS-105 and KBS-107.

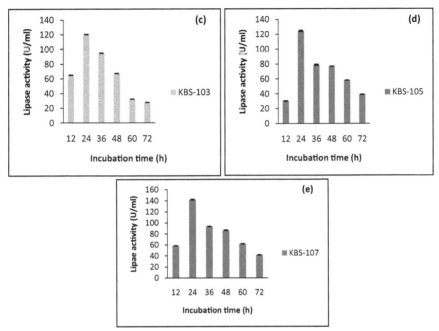

**FIGURE 10.19(a)–(e)**   *(Continued)*

## 10.8.4   EFFECT OF AGITATION RATE

The culture flasks are agitated in a temperature controlled orbital shaking incubator in the range of 50–300 rpm, there after the production of lipase monitored. Different agitation speeds were tested for lipase production by the bacterial strains and the optimum production by the strains KBS-101, KB2F, KBS-103, KBS-105 and KBS-107 was at 200 rpm (Fig. 10.20).

All proteins suffer denaturation, and hence loss of catalytic activity with increase in time. The maximum lipase production from the strains KBS-101, KBS-103, KBS-105 and KBS-107 was achieved after 24 h of incubation at 37°C whereas KB2F after 48 h at 45°C. As stated by Ibrahim (1994) the aeration rate in the stirred tank bioreactor is important for the growth of the bacteria, and also may enhance enzyme production. Rao et al. (1993) observed that agitation rate in a stirred tank bioreactor plays an important role in the production of thermostable lipase enzyme and is an important parameter to ensure nutrient availability in a growth medium having lipid substrates. The maximum accumulation of lipase

was obtained from all culturable bacterial strains at 200 rpm. Satnayarayana (1994) reported that higher agitation also increases gas dispersion,
and increased gas dispersion allows more mass transfer.

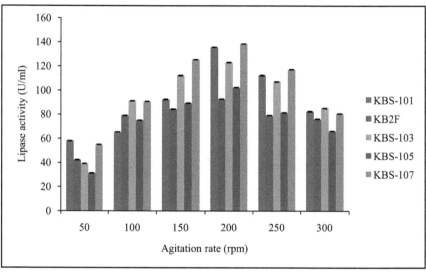

**FIGURE 10.20**   Effect of agitation on lipase production by bacterial strains under
submerged fermentation.

### 10.9.5   EFFECT OF SUBSTRATE

To determine the substrate specificity of culture supernatant containing
crude lipases from bacterial strains KBS-101, KB2F, KBS-103, KBS-105
and KBS-107, substrates like tributyrin, olive oil, soybean oil, sunflower
oil, palm oil, coconut oil and castor oil are used at the final concentration
of 1% (v/v) and incubated at 37°C for 30 min. The lipolytic activity is
assayed by the titrimetric method (Fig. 10.21(a)–(e)).

The maximum lipase production was obtained from the bacterial
strains KBS-101, KBS-103 and KBS-105 in the presence of 1.0% olive oil
substrate in the culture medium; while strains KB2F and KBS-107 showed
maximum lipase production in the presence of 1.0% tributyrin.

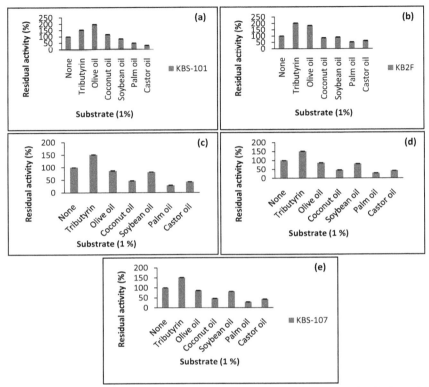

**FIGURE 10.21(a)–(e)**   Effect of various substrates on lipase production by the bacterial Strains KBS-101, KB2F, KBS-103, KBS-105 and KBS-107. The activity of lipases measured without any substrate was taken as 100%.

### 10.9.6   EFFECT OF METAL IONS

To investigate the effect of the different divalent cations on lipase activity, the titrametric assay is performed with crude enzyme from bacterial strains KBS-101, KB2F, KBS-103, KBS-105 and KBS-107 in the presence metal ions like $Li^{2+}$, $Ca^{2+}$, $Ni^{2+}$, $Mg^{2+}$, $Hg^{2+}$, $Co^{2+}$, $Cu^{2+}$, $Fe^{2+}$, $Zn^{2+}$, $Mn^{2+}$ and $Cd^{2+}$. The enzyme activity without metal ions serves as the control and is considered as 100%.

The metal ions of $Ca^{2+}$ enhanced the activity of lipase produce by KBS-101 and KBS-103, $Fe^{2+}$ enhanced the activity of KB2F and KBS-107, $Co^{2+}$ of KBS-105 and $Mg^{2+}$ of KBS-107. Metal ions $Li^{2+}$, $Ni^{2+}$, $Cu^{2+}$ and $Cd^{2+}$ showed inhibitory effect on the activity of lipase produced by KBS-101, KB2F, KBS-103, KBS-105 and KBS-107. Data obtained are presented in Figure 10.22(a)–(e).

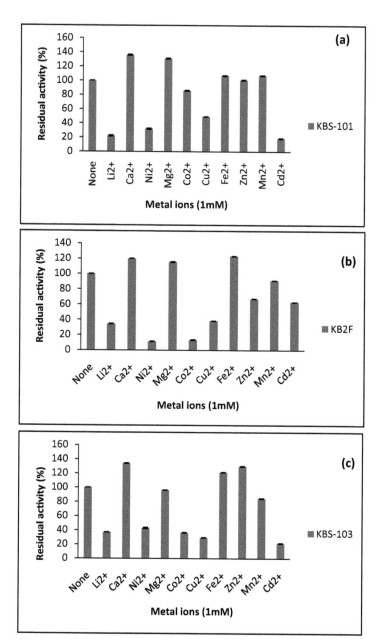

**FIGURE 10.22(a)–(e)** Effect of metal ions on the activity of crude lipase produced by the bacter strains KBS-101, KB2F, KBS-103, KBS-105 and KBS-107. The activity of lipases measured withc any metal ions was taken as 100%.

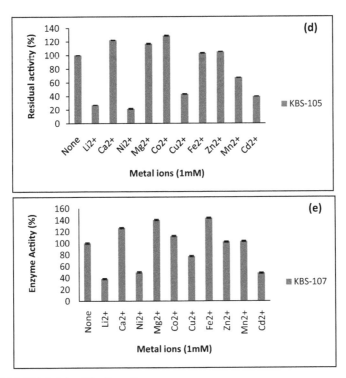

**FIGURE 10.22(a)–(e)**    (Continued)

Out of six lipid sources at 1.0%, maximum lipase production from KBS-101, KBS-103 and KBS-105 was obtained in the presence of 1% olive oil, KB2F and KBS-107 in tributyrin. Bora and Kalita (2012) reported that lipase productivity by *Bacillus* sp. DI14 strain was higher when cultured in medium supplied with vegetable oils. The present finding is in agreement with the inducible effects of olive oil and tributyrin on lipase production by *Rhizopus* sp. BTNT-2 as described by Bapiraju et al. (2005). According to Matsumae et al. (1994) the effect of metal ions could be attributed to a change in the solubility and the behaviour of the ionized fatty acids at inter-faces, and from a change in the catalytic properties of the enzyme. Out of the divalent metal ions tried, $Li^{2+}$, $Ni^{2+}$, $Cd^{2+}$, $Cu^{2+}$ and $Mn^{2+}$ salts reduced the activity of lipases. The inhibitory effect of the heavy metal $Cu^{2+}$ ion on lipases was similar to that of *Pseudomonas aeruginosa* (Dharmsthitiand Kuhasuntisuk, 1998), *Pseudomonas fluorescens* (Makhzoum et al., 1996). Metal ions $Ca^{2+}$ enhanced the activity of produced by KBS-101 and KBS-103, $Fe^{2+}$ in KB2F, $Co^{2+}$ in KBS-105 and $Mg^{2+}$ in KBS-107. Salts of heavy metals such as $Cu^{2+}$ strongly inhibited the lipase, suggesting its involvement in the alteration of enzyme conformation.

## REFERENCES

Ananthi, S.; Immanuel, G.; Palavesam, A. Optimization of Lipase Production by *Bacillus Cereus* Strain Msu as Through Submerged Fermentation. *Plant Sci. Feed.* **2013,** *3,* 31–39.

Bapiraju, K.V.V.S.N., et al. Sequential Parametric Optimization of Lipase Production by a Mutant Strain *Rhizopus* sp. BTNT-2. *Bra. J. Chem. Eng.* **2005,** 45, 257–273.

Bonala, K. C.; Mangamoori, L. N. Production and Optimization of Lipase from *Bacillus Tequilensis* Nrrl B-41771. *Int. J. Biotechnol. Appl.* **2012,** *4,* 134–136.

Bora, L.; Bora, M. Optimization of Extracellular Thermophilic Highly Alkaline Lipase from Thermophilic *Bacillus sp* Isolated from Hotspring of Arunachal Pradesh, India. *Braz. J Microbiol.* **2012,** *43,* 30–42.

Dharmsthiti, S.; Kuhasuntisuk, B. Lipase from *Pseudomonas Aeruginosa* LP602: Biochemical Properties Application for Wastewater Treatment. *J. Ind. Microbiol. Biotechnol.* **1998,** *21,* 75–80.

Gupta, R.; Gupta, N.; Rathi, P. Bacterial lipases: An Overview of Production, Purification and Biochemical Properties. *Appl. Microbiol. Biotechnol.* **2004,** *64,* 763–781.

Ibrahim, C. O. *Enzyme Sumber Mikrob.* In: Pengantar Mikrobiologi Industri. Universiti Sains: Malaysia; pp 73–130, 1994.

Ito, T., et al. Lipase Production in Two-Step Fed-Batch Culture of Organic Solvent-Tolerant Pseudomonas Aeruginosa LST-03. *J. Biosci. Bioeng.* **2001,** *91,* 245–250.

Makhzoum, A.; Owusu-Apenten, R. K.; Knapp, J. S. Purification and Properties of Lipase from *Pseudomonas fluorescens* strain 2D. *Int. Diary J.* **1996,** *6,* 459–472.

Matsumae, H.; Shibatani, T. Purification and Characterization of Lipase from *Serratia Marcescens* Sr 41 8000 Responsible for Asymmetric Hydrolysis of 3- Phenylglycidic Esters. *J. Ferment Bioeng.* **1994,** *77,* 152–158.

Parr, J. F.; Sikora, L. J.; Burge, W. D. Factors Affecting the Degradation and Inactivation of Waste Constituents in Soil, *in Land Treatment of Hazardous Wastes.* J. F. Parr et al., Eds.; Noyes Data Corp., park Ridge: NJ, 321337, 1983.

Rao, P. V.; Jayaraman, K.; Lakshmann, C. M. Production of Lipase by Candida Rugosa in Solid State Fermentation. 1: Determenation of Significant Process Variables. *Proc. Biochem.* **1993,** *28,* 385–389.

Sambrook. J.; Russel, D. W. *Molecular Cloning: A Laboratory Manual,* 3rd ed.; Cold Spring Harbor Laboratory Press: New York, 2001.

Saitou, N.; Nei, M. The Neighbour-Joining Method: a New Method for Reconstructiong Phylogenetic Trees. *Mol. Biol. Evol.* **1987,** *4,* 406–425.

Satnayarayana, T. *Production of Extracellular Enzymes by Solid State Fermentation.* In solid state fermentation; Pandey A, Ed.; Wiley Eastern Ltd.: New Delhi, India, 1994; 122–129.

# CHAPTER 11

# MICROBIAL ASSAY OF CULTURE SUPERNATANT CONTAINING CRUDE LIPASE

## CONTENTS

### 11.1 DETERMINATION OF ANTIBACTERIAL ACTIVITY

The lipase-producing bacterial strains are subjected to the screening for antibacterial activity using different fungal strains.

#### 11.1.1 TEST MICROBES

The bacterial strains used in the present investigation are obtained from our own laboratory, (Department of Molecular Biology and Biotechnology, Tezpur, Assam, India). The strains are *Escherichia coli* (MTCC 40), *Staphylococcus aureus* (MTCC 737), *Klebsiella pneumoniae* (MTCC 109), *Bacillus subtilis* (MTCC 619) and *Pseudomonas aeruginosa* (MTCC 7815).

Antibacterial activity of lipase-producing bacterial strains is evaluated by the well diffusion method. Stock cultures are maintained at 4°C on Mueller-Hinton agar (MHA) medium. Active cultures are prepared by transferring a loopful of bacterial cells from each stock culture to the test tubes containing sterilized Mueller-Hinton broth (MHB) and incubating at 37°C for 24 h in a rotary shaker.

The MHA plates are prepared by pouring 20 ml of molten media into the sterile Petri dishes. The plates are allowed to solidify for 20 min and then 100 µl of the test bacteria in the log phase of growth ($10^6$–$10^8$ cells as per Mc Farland standard) are spread on the surface of MHA medium using a sterile glass spreader. Wells having diameter of 6.0 mm are made on MHA plates using a sterile corkborer. 50.0 µl of each lipase-producing bacterial culture is introduced into the separate wells. Streptomycin sulphate ($1 \text{ mg·ml}^{-1}$) 50.0 µl, which is a broad spectrum antibiotic, is used as the positive and MHB as the negative control. After overnight incubation of plates at 37°C, the growth of bacteria is determined by measuring the diameter of inhibition zone using a transparent metric ruler (Himedia, India).

The culture supernatant containing crude lipases isolated from the bacterial strains KBS-101, KB2F, KBS-103, KBS-105 and KBS-107 were subjected to antibacterial activity test and and data obtained are presented in Table 11.1 the same are shown in Figure 11.1.

**TABLE 11.1**  Antibacterial Assay of Culture Supernatant Containing Crude Lipases from Bacterial Strains KBS-101, KB2F, KBS-103, KBS-105 and KBS-107.

| Bacterial strains | Zone of inhibition (mm) | | | | |
|---|---|---|---|---|---|
| | *Escherichia coli* (MTCC 40) | *Bacillus subtilis* (MTCC 619) | *Staphylococcus aureus* (MTCC 737) | *Pseudomonas aerugenosa* (MTCC 7815) | *Klebsiella pneumonae* (MTCC 109) |
| KBS-101 | 12 | 11 | 13 | 10 | 14 |
| KB2F | – | – | – | – | – |
| KBS-103 | 13 | 10 | 11 | – | – |
| KBS-105 | 10 | – | – | – | – |
| KBS-107 | – | – | – | – | – |

The culture supernatant containing crude lipase from KBS-101and KBS-103 showed strong antibacterial activity against both Gram-positive and Gram-negative bacteria. However, the culture supernatant containing crude lipase from KB2F and KBS-107 failed to show inhibitory effect on both Gram-positive and Gram-negative bacterial species; KBS-105 showed antibacterial activity against the Gram-negative bacteria such as *E. coli*. The maximum zone of inhibition was observed in the case of KBS-101 against *K. pneumonae* followed by *S. aureus* and *E. coli*. The standard antibacterial drug streptomycin (1 mg/ml) was used as the positive control for the experiment.

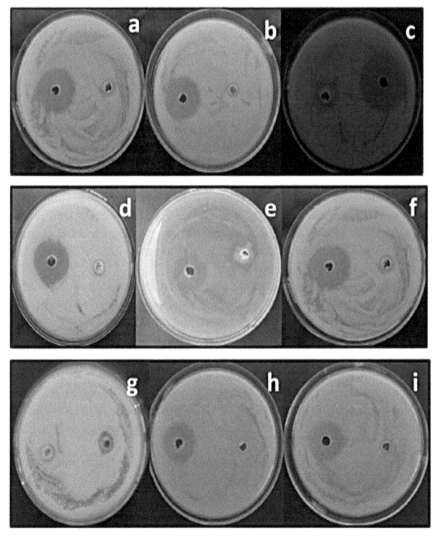

**FIGURE 11.1**    Antibacterial activity of crude lipase isolated from KBS-101 against (a) *Bacillus subtilis* MTCC 619 (b) *Pseudomonas aeruginosa* MTCC 7815 (c) *Staphylococcus aureus* MTCC 737 (d) *Escherichia coli* MTCC 40 (e) *Klebsiella pneumonia* MTCC 109, KBS-103 against (f) *E. coli* MTCC 40 (g) *S. aureus* MTCC 737 (h) *B. subtilis* MTCC 619 and from KBS-105 against (i) *E. coli* MTCC 40.

## 11.2   DETERMINATION OF ANTIFUNGAL ACTIVITY

The lipase-producing bacterial strains are subjected to screening for anti-fungal activity using the different fungal strains.

### 11.2.1   TEST ORGANISMS

The fungal strains used in the present investigation are obtained from our own laboratory, (Department of Molecular Biology and Biotechnology, Tezpur, Assam, India). The fungal strains are *Candida albicans* (MTCC 227) and *Fusarium oxysporum* (MTCC 284).

Antifungal activity of lipase-producing bacterial strains is evaluated by the well diffusion method. Fungal cultures are maintained at room temperature in potato dextrose agar (PDA) medium. Active cultures of the fungal strains are prepared by seeding a loopful of fungi into potato dextrose broth (PDB) and incubating without agitation for 48 h at room temperature. The culture is diluted with PDB to achieve the optical density corresponding to $2.0 \times 10^5$ spores/ml.

Potato dextrose agar plates are prepared by pouring 20 ml of molten medium into the sterile Petri dishes. The plates are allowed to solidify for 20 min and then 100 µl of the test fungal strains are spread on the surface of the PDA medium using a sterile glass spreader. With the help of a sterile corkborer wells having 6 mm diameter each are made on PDA plates. 50 µl each of cultured lipase-producing bacteria are introduced into one of the wells. Amphotericin ($1$ mg·ml$^{-1}$) is taken as the positive control. After the overnight incubation of plates at 37°C, fungal growth is determined by measuring the diameter of inhibition zone using a transparent metric ruler (Himedia, India).

Antifungal activity of the culture supernatant containing crude lipase isolated from the bacterial strains KBS-101, KB2F, KBS-103, KBS-105 and KBS-107 was assessed against *C. albicans* (MTCC 227) and *F. oxysporum* (MTCC284). The culture supernatant containing crude lipase from KBS-101 showed zone of inhibition against *F. oxysporum* (MTCC284) with a diameter of 11.0 mm. The antifungal assay data obtained are presented in Table 11.2.

**TABLE 11.2**   Antifungal Assay of the Crude Lipases from the Bacterial Strains KBS-101, KB2F, KBS-103, KBS-105 and KBS-107.

| Bacterial Strains | Zone of inhibition (mm) | |
|---|---|---|
| | *Candida albicans* (**MTCC 227**) | *Fusarium oxysporum* (**MTCC 284**) |
| KBS-101 | – | 18 |
| KB2F | – | – |
| KBS-103 | – | – |
| KBS-105 | – | 11 |
| KBS-107 | – | – |

Antimicrobial peptides (AMPs) are a unique, diverse group of molecules usually synthesized by microorganisms as well as multicellular organisms and are part of the innate host defense system. As stated by Zasloff (2002), the treatment of animals with antibiotics leads to antibiotic residue problem in veterinary products and possible transmission of antibiotic-resistant microbes to humans. Antimicrobial peptides are ubiquitous in nature, with high selectivity against target organisms and resistance against them is rarely observed. On the basis these limitations of commercially available antimicrobial agents used in food industries, animal husbandry and agriculture the antibacterial property of culture supernatant containing crude lipases was assayed.

The culture supernatant containing crude lipases from the culturable bacterial strain KBS-101 inhibited growth of both Gram-positive and Gram-negative bacteria. KBS-101 lipase exhibited maximum inhibitory effect against *K. Pneumonia (*MTCC109) followed by *S. aureus* (MTCC 737), *E. coli* (MTCC 40), *B. subtilis* (MTCC 619) and *Pseudomonas aerugenosa* (MTCC 7815). KBS-103 exhibited inhibitory effect against *E. coli* (MTCC 40), *B. subtilis* (MTCC 619) and *S. aureus* (MTCC 737), and KBS-105 against *E. coli* (MTCC 40). Wu et al. (2013) reported similar results by working on bacterial species/strains. The bacterial strain KBS-105 could not bind lipoteichoic acid (LTA) hence the growth of Gram-positive bacteria could not be inhibited by the lipase of this strain. Lipase of KB2F and KBS-107 did not exhibit antibacterial activity against Gram-positive and Gram-negative bacteria.

In the case of Gram-negative bacteria, peptidoglycan is not exposed on the cell surface since it is located underneath the outer membrane containing lipopolysaccharides (LPS). Therefore, killing of Gram-negative bacteria most likely involves initial binding of AMPs (Mahamoud et al., 2007). In Gram-positive bacteria, the initial binding of AMPs may involve LTA in addition to peptidoglycan because LTA also inhibits killing of bacteria by AMPs (Nasir et al., 2010). Considering its antibacterial, biochemical and pharmacological properties its safe application as preservative for preventing or slowing microbial growth, which is the major reason of spoilage and poisoning of food products, is warranted. Antimicrobial peptides are advantageous option for use as new preservatives to prevent the problems that are encountered due to irregular use of conventional antibiotics. The peptides kill Gram-negative and Gram-positive bacteria and may also have the ability to enhance immunity by functioning as immunomodulators. According to Bechinger (1997), the killing mechanisms of most peptides consist of attack on the outer and inner membranes, ultimately resulting in lysis of the bacteria. The cytoplasmic membrane is a frequent target, but peptides may also interfere with DNA and protein synthesis, protein folding and cell wall synthesis. Alternately, they may penetrate into the cell to bind intracellular molecules which are crucial to cell living. Intracellular binding includes inhibition of cell wall synthesis, alteration of the cytoplasmic membrane, activation of autolysin, and inhibition of DNA, RNA and protein synthesis.

The culture supernatant containing crude lipase from KBS-101 and KBS-105 showed good antifungal activity against *F. oxysporum* (MTCC 284). Saha et al. (2012) reported similar results in the case of *Bacillus* sp. In the present investigation, crude lipases from KBS-101 and KBS-105 inhibited the growth of fungal strains which might degrade the fungal cell wall and the same was also supported by Shali et al. (2010)

## REFERENCES

Bechinger, B. Structure and Functions of Channel-Forming Peptides: Magainins, Cecropins, Melittin and Alamethicin. *J. Membr. Biol.* **1997,** *156,* 197–211.

Mahamoud, A., et al. Antibiotic Efflux Pumps in Gram-Negative Bacteria: The Inhibitor Response Strategy. *J. Antimicrob. Chemother.* **2007,** *59,* 1223–1229.

Nasir, N. A., et al. Antibacterial Properties of Tualang Honey and its Effect in Burn Wound Management: A Comparative Study. *BMC Complementary Altern. Med.* **2010,** *10,* 1–7.

Saha, D., et al. Isolation and Characterization of Two New *Bacillus subtilis* Strains from the Rhizosphere of Eggplant as Potential Biocontrol Agents. *J. Plant Pathol.* **2012,** *94,* 109–118.

Shali, A., et al. *Bacillus pumilus* SG2 Chitinases Induced and Regulated by Chitin, Show Inhibitory Activity Against Fusarium Graminearum and Bipolaris Sorokiniana. *Phytoparasitica*. **2010,** *38,* 141–147.

Wu, W. J.; Park, S. M.; Ahn B. Y. Isolation and Characterization of an Antimicrobial Substance from *Bacillus subtilis* BY08 Antagonistic to *Bacillus cereus* and *Listeria monocytogenes*. *Food Sci. Biotechnol*. **2013,** *22,* 433–440.

Zasloff, M. Antimicrobial Peptides of Multicellular Organisms. *Nature*. **2002,** *415,* 389–395.

# CHAPTER 12

# CRITICAL OBSERVATIONS

Sequence-based metagenomic study for the assessment of microbial diversity followed by functional approach for the production of lipase/esterase enzyme system and genomic study of lipase-producing culturable bacteria lead to following conclusions:

1. A simple and suitable method is developed for the isolation of soil metagenomic DNA (mgDNA). NaCl, Sodium dodecyl sulphate (SDS) and heating at 72°C temperature to lyse the bacterial cells followed by isopropanol precipitation without using phenol cleanup. The yield of mgDNA is 3.8 µg/g soil with the purity of 1.6 and 0.8 with respect to protein and humic acid, respectively. Further purification of the extracted mgDNA using Sephadex G-50 column (MP1 method) has yielded higher amount of DNA suitable for further molecular analyses. Therefore, DNA extracted using the method M5 (reduction in humic acid), followed by purification (MP1) is successfully used for the construction of mgDNA library.

2. The metagenomic study of microbial diversity present in a bakery soil sample has revealed three 16S rRNA gene clones which are named as KBSR1, KBSR2 and KBSR3 with sequence-based analysis. Subsequently, the 16S rDNA gene sequences are deposited in GenBank of NCBI with accession numbers KJ685805, KJ700876 and KJ700877 respectively. 16S rRNA gene clones with accession numbers KJ685805 (KBSR1), entries KJ700876 (KBSR2) and KJ700877 (KBSR3) have represented 97–99% sequence homology with those of unculturable bacteria. The present study has confirmed that 99% microorganisms in a particular soil sample are unculturable.

3. The metagenomic library is constructed into pET-32a vector with the insert size of 0.5–3.0 kb from the soil metagenome. The functional metagenomic approach for the lipase enzyme coding gene from soil metagenome has revealed a positive clone KBS-plip1,

showing lipolytic activity after screening in tributyrin (1%) agar medium and the size of the lipase coding gene is 891 bp. Subsequently, the nucleotide sequence of lipase coding gene is deposited in GenBank of NCBI data library and obtained the accession number KF743145 for the uncultured bacterium KBS-plip1. The sequence homology of KBS-plip1 has represented 84–99% identity with that of the uncultured bacterium.

4. The expression study of KBS-plip1 has represented the overexpression of the cloned gene (891 bp) in the presence of 0.5 mM isopropyl β-D-1-thiogalactopyranoside (IPTG) for 5 h at 30°C. The purification of the expressed protein using Ni(II)–nitrilotriacetic acid (Ni–NTA) affinity chromatography has revealed a single band of protein approximately 40 kDa in size which is quite higher than that of the expected molecular weight of KBS-plip1 that is 32,560 kDa. However, the increased molecular weight of the protein could be due to the fusion of His-tag (12.7 kDa) with the expressed protein.

5. The nucleotide sequence of KBS-plip1 on translation to the protein sequence (ExPASy translation tool) has exhibited the presence of 296 amino acids. The BLAST result of the amino acid sequence of KBS-plip1 has revealed a maximum of 55% sequence homology with the lipase enzyme (3V9A) produced by the uncultured bacterium.

6. The multiple sequence alignment of KBS-plip1 has showed the presence of conserved HGGG motif (amino acids 23–26) upstream of the active site, catalytic consensus pentapeptide GDSAG, catalytic triad DPM (amino acids 145–147) and HVF (amino acids 228–230) referring to the conserved sequences of hormone-sensitive lipase (HSL) family. This confirms that KBS-plip1 belongs to the HSL family. KBS-plip1 contains GDAAG in place of GDSAG and it could be the novel functional characteristic for the lipase enzyme.

7. Homology models and validation for protein structure prediction for KBS-plip1 have revealed 72% structural similarity with that of ABZ9 template and showed a generated ribbon model in a dimmer fold, consisting of nine β-strands attached with seven α-helices. The Ramachandran plot for modelled enzyme has consisted of favoured (95.6%), allowed (4%) and outlier (0.4%). The result

is significant as a high percentage of residues is in the favoured region (>95%). This indicates that the built in model is of good quality.

8. Biochemical characterization of KBS-plip1 for dose-dependent study has showed the required concentration of 1.0 $\mu g \cdot ml^{-1}$ for maximum enzyme activity in the presence of 1% tributyrin. The enzyme kinetics for KBS-plip1 has revealed $V_{max}$ and $K_m$ values for the enzyme catalysed reaction to be 227 U/mg and 0.0806 mg/ml, respectively.

9. The optimum pH and temperature for the maximum enzyme activity of KBS-plip1 is found to be 7.5 and 37°C, respectively. It has showed maximum lipase activity in the presence of 0.5% CTAB, 0.5% gum arabic and 1 mM EDTA. Metal ions $Ca^{2+}$, $Mn^{2+}$, $Fe^{2+}$ and $Zn^{2+}$ in the final concentration of 1 mM and salt NaCl at 1.5 M concentration have enhanced the enzyme activity of KBSplip1 by two fold.

10. KBS-plip1 has exhibited solvent tolerance activity and revealed maximum activity in the presence of 30% of 2-propanol. Solvents like ethanol, 1-propanol, glycerol, acetone, acetonitrile and DMSO have also enhanced the enzyme activity.

11. The KBS-plip1 derived lipase enzyme has revealed its stability at 37°C for 100 min and the same decreased drastically between 45 and 55°C with half-life of 60 and 20 min, respectively.

12. Five different lipase producing culturable bacterial isolates KBS-101, KB2F, KBS-103, KBS-105 and KBS-107 are isolated from a bakery waste deposited soil, Tezpur, Assam after screening in tributyrin agar plate medium. All five bacterial strains have exhibited zone of hydrolysis 28–25 mm. Morphological and biochemical screening of the lipase producing bacterial strains have showed the strains KBS-101, KB2F and KBS-105 to be rod shaped gram positive *Bacillus* sp.; KBS-103 and KBS-107 rod shaped Gram-negative *Enterobacter* sp.

13. The culturable bacterial strains KBS-101, KB2F, KBS-103, KBS-105 and KBS-107 have yielded genomic DNA 73.4, 58.8, 44.1, 33.9 and 72.2 µg/ml, respectively with the purity of 1.85–1.94.

14. The lipase producing culturable bacterial strain KBS-101 has showed maximum growth at 37°C during 16 h, KBS-102 at 45°C

during 24 h, KBS-103 at 37°C during 16 h, KBS-105 at 37°C during 24 h and KBS-107 at 37°C during 24 h of incubation.

15. The crude bacterial lipases extracted from the bacterial strains have exhibited high degree of thermal stability, and retained 50% of their catalytic activity at 60°C after 80 min. Compatibility study of the culture supernatant containing crude lipases has exhibited higher activity with the commercial detergents. The culture supernatant containing crude lipases have retained their 50% activity even after 30 days of storage at 4°C.

16. An inoculum size of 2.0% gives maximum lipase production. Glucose for the strains KBS-101 and KBS-105; lactose for KB2F; and galactose for KBS-103 and KBS-107 are the effective carbon source for maximum lipase production. Beef extract for KBS-101 and KB2F, yeast extract for KBS-103 and KBS-107 and peptone for the strain KBS-105 are good nitrogen sources for maximum lipase production.

17. KBS-101 and KBS-103 have showed maximum lipase production after 24 h with maximum cell dry biomass; but strains KB2F, KBS-105 and KBS-107 showed maximum lipase production after 48 h with the highest cell dry biomass.

18. The optimum pH for maximum lipase production by KBS-101 and KBS-105 is 8.5, KB2F and KBS-103 at 7.5, whereas KBS-107 at 9.0 pH. The optimum temperature for maximum lipase production by strains KBS-101, KBS-103, KBS-105 and KBS-107 is 37°C and by KB2F 45°C. The maximum lipase production from KBS-101, KBS-103, KBS-105 and KBS-107 is achieved after 24 h of culture at 37°C while KB2F after 48 h of culture at 45°C. The optimum agitation rate for maximum lipase production from all five bacterial strains is 200 rpm.

19. Olive oil is found to be a good substrate for maximum lipase production by the bacterial strains KBS-101, KBS-103 and KBS-105, and tributyrin for the highest lipase production by KB2F and KBS-107.

20. The divalent metal ion $Ca^{2+}$ enhanced the activity of the culture supernatant containing crude lipase enzyme extracted from KBS-101 and KBS-103; $Fe^{2+}$ from KB2F and KBS-107, and $Co^{2+}$ from KBS-105.

21. The culture supernatant containing crude lipase from the bacterial strain KBS-101 has showed the highest antibacterial activity against Gram-positive and Gram-negative bacteria; and antifungal activity against *Fusarium oxysporum* (MTCC 284).

# INDEX